T5-CPW-919

BIOLOGICAL RESEARCH METHOD

Practical Statistics for Non-mathematicians

BIOLOGICAL RESEARCH METHOD

Practical Statistics for Non-mathematicians

H. H. HOLMAN
D.Sc., Ph.D., F.R.C.V.S.
Formerly Deputy Director, Institute for Research on Animal Diseases Compton, Berks

SECOND EDITION

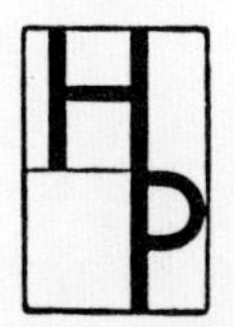

HAFNER PUBLISHING COMPANY
NEW YORK

Published in Great Britain by
OLIVER AND BOYD LTD
Edinburgh

First edition 1962
Second edition 1969

Printed in Great Britain by
Thomas Nelson (Printers) Ltd. London and Edinburgh

PREFACE

This book is intended as a personal guide for research workers who carry out experiments involving animals or their products.

Without guidance a scientist, although giving full rein to his imagination, may neglect the main factor that will build his reputation. This factor is reliability; without this his work will never be spoken of with respect. It is important to acquire reliability early, for a few years spent on experiments that lack adequate controls will lead to an ill-disciplined technique which is hard to remedy. This book offers guidance on the design, recording and analysis of experiments according to rules used in realistic research.

One of the tools of the trained research worker is biometrics because, even if his data are analysed by a professional statistician, he still needs to know the implications that can be drawn from these measurements, how his data should be displayed, and the logic of experimental design. This book introduces these subjects and it also offers the reader a selection of simple tests from which he may choose those that suit his purpose.

PREFACE TO SECOND EDITION

In this new edition there have been no changes in the advice given on carrying out experimental work. The statistical side remains utilitarian and is based on my belief that both examiners for higher degrees and scrutineers of scientific papers prefer a simple test that the writers can discuss rather than a more sensitive but intricate test which, when challenged, elicits the reply that the writer is not responsible as it was done by a friend. With this in mind, and emboldened by Dr D. R. Cox's book "The Planning of Experiments", I have now introduced examples using the index of response and randomised blocks as substitutes for the analysis of covariance. I have found it hard to obtain information on multiple samples from standard textbooks and have therefore tried to place more emphasis on their use, including the calculation of a regression by a method kindly given me by Dr R. C. Campbell. I have augmented the section on nonparametric tests by including details of the corner test, and through the kindness of Dr Colin White and the editors of *Biometrics* I have included the greater part of the tables for Dr White's test.

ACKNOWLEDGEMENTS

I would like to express my thanks to I. H. Pattison, B.SC., M.R.C.V.S., and to G. B. S. Heath, B.SC., F.R.C.V.S., for their criticism of the first draft and for their very kind encouragement. My task was made easier by having worked and exchanged ideas with Pattison over many years and I am also obliged to him for having read the proofs.

In trying to understand and explain the basic terms and rationale of statistical methods I am very indebted to my brother, Lt.-Col. L. J. Holman, B.SC., F.B.PS.S., for answering, patiently and with lucidity, many questions over a large number of years.

The abridged Tables III, IV and V are from the statistical tables of Fisher and Yates detailed in the bibliography, and I am grateful for the authors' permission to publish them. I also wish to thank Prof. D. Mainland for his permission to use part of his Table III from the binomial tables detailed in the references and in the bibliography.

There is little that is original in the text except, perhaps, in the undetected errors, and I would be most grateful if readers would point these out to me.

H. H. HOLMAN

CONTENTS

NOTES

Tables and Figures take their numbers from the pages on which they appear. Specific references are detailed at the end of each chapter. General references are to books in the Selected Bibliography

CHAPTER 1

TRUTH, LOGIC AND CHANCE

> Man knows that, at any moment, he can tell a lie that, for a while will delay or divert the working of cause and effect. Being an animal who is still learning to reason, he does not yet understand why, with a little more, or a little louder, lying, he should not be able permanently to break the chain of that law.
>
> RUDYARD KIPLING

A biologist engaged in research work should be able to carry out a workmanlike experiment. A workmanlike experiment is one that is well planned, well carried out, and well recorded. As a result the experimental data are easy to assess and the experiment is easy to write up. Such a technique is best learnt from personal supervision but this book attempts to show the beginner some of the basic principles needed to produce reliable and definite results. It should also help the beginner to avoid those peculiar indefinite results which still appear in some journals and which read something like this: " The experiment does not permit us to say definitely that the addition of this new substance to the food always produced an advantageous effect, but animals number 7 and number 11 appeared to show great improvement; and although there was some loss in number 3 this perhaps could be accounted for by assuming that this animal was much older than the others."

It is also hoped that it will encourage readers to avoid results which, if truthfully recorded, might read like this: " It appears likely that there was some rhythm present in these measurements, but it was not possible to prove this because the writer always liked to leave the laboratory at 5 o'clock and made no measurements on Saturdays and Sundays."

TRUTH

Scientific experiments are carried out on objects and animals in order to establish the truth. Most people believe it is quite easy to establish the truth but the old adage states that " Truth lies at the bottom of the well." Things lying at the bottom of a well are difficult to get up, in fact when a dead cat was reported to be lying at the bottom of one of the local wells, all the authorities did was to put up a notice saying that the water was unfit for human consumption. Whether truth is fit for human consumption is a debatable point, but certainly there are more people concerned in its suppression than there are in its production. In trying to proclaim the truth, science is often in conflict not only with politics, religions and commerce but with individual scientists.

However, it is still the grasping of the truth that is the most difficult task, and the unfortunate thing is that this appears to be an easy thing to do and we all feel that we are particularly good at it. Yet the fact remains that when senior research workers are told that recent work has shed new light on a subject, the question immediately asked is, not what method was used, or where it was done, but who did it? The work of one man will be accepted immediately; the work of another will not be, however low the probability of his statistical analysis. The fact that after years of doing research work some workers can still produce results that are suspect is sufficiently interesting to warrant a little closer look at truth.

The first thing to recognise is that truth is not an instinct. On the contrary, in all but the highest civilisations, man's instincts for preservation are served better by his ability to deceive. In trying to explain the nature of truth Trotter (1916), postulated that with gregarious animals there was a " herd instinct " which made the individual instinctively suggestible to the wishes of the herd. In man this instinct provided such emotions as patriotism, religion, esprit de corps and racial and class prejudices. The suggestibility of this instinct was in some cases so powerful that it could overcome even the individual instinct of self-preservation.

Thus, where our experimental findings appear to support the belief or opinion of our group, then conscience is not a reliable guide to the truth, and we tend to accept the result without question, in fact, to quote Trotter:

" When, therefore, we find ourselves entertaining an opinion about the basis of which there is a quality of feeling which tells us that to inquire into it would be absurd, obviously unnecessary, unprofitable, undesirable, bad form, or wicked, we may know that that opinion is a non-rational one, and probably, therefore, founded upon inadequate evidence."

If this is the case, how can we ever get to the state where we wish to know the truth for its own sake? Trotter suggests that "The solution would seem to lie in seeing to it that suggestion always acts on the side of reason; if rationality were once to become really respectable, if we feared the entertaining of an unverifiable opinion with the warmth with which we fear using the wrong implement at the dinner table, if the thought of holding a prejudice disgusted us as does a foul disease, then the dangers of man's suggestibility would be turned into advantages. We have seen that suggestion already has begun to act on the side of reason in some small part of the life of the student of science, and it is possible that a highly prophetic imagination might detect here a germ of future changes."

This suggestibility must be sought by acquiring a liberal education and by mixing with other truth seekers in universities or as a member of scientific societies. It is only when the scientific method of "truth by verification" has become a matter of conscience that a research worker can regularly design the worth while experiment which is intended, not to bolster up an opinion that he feels will bring him fame because of its value to the community, but to test it in such a way that, where necessary, it will show his opinion to be completely untrue.

This attitude of mind is the most important factor in good research work, but it is not sufficient in itself, for good experimentation is both a discipline and an art. And even men who, through education and shrewd observation, have attained eminence in a biological science, can produce pitiably inadequate experiments if they lack training in experimental method.

Art cannot be taught, it can only be demonstrated. It is therefore fortunate that the greater the art of the experimenter the more simple and direct will be his experiments, so that in acquiring the simple methods of logic and biometrics mentioned in this book the novice is learning to use the tools of the expert.

Logic

Second to the desire for truth comes the need for a reliable method of reasoning. The process of reasoning that we are accustomed to use in our daily life is that referred to as " Post hoc ergo propter hoc."—after that therefore because of that—and this method is based on the fact that the cause must precede the effect. Thus if a man and a dog are nearly dying from thirst and they reach a stagnant pond, the man may hang back and let the dog drink the water first. If after drinking the dog runs round in circles and then falls over dead, it is commonsense for the man to accept the " post hoc " argument that the water is poisonous; if the man is an anti-vivisectionist he will keep the dog back and drink the water himself first, it is then up to the dog to apply a "post hoc" argument.

Although the "post hoc" argument plays a large and useful part in everyday affairs, it can be very misleading. The man who, after a night out in which he has mixed every sort of drink, comes home feeling sick and tells his wife that it must have been something he ate, has probably chosen the wrong antecedent as the cause, whereas the mother who explains that her little girl has lovely teeth, curly hair, and never has a cold, because she has always been fed on wholemeal bread, may not have selected the right attributes as effects.

In experiments on animals, in addition to these wrong conclusions based on faith or self-deception, mistaken interpretation may be due to the natural waxing and waning of physiological or disease processes, so that the " post hoc " argument becomes quite useless as a weapon in the search for truth. Nevertheless it must be admitted that in spite of this there are committees that even today, permit observations of this type in which the only thing that is certain is that the result will be debatable. The use of such an inefficient method of inference in experiments involving domesticated animals is often encouraged by the owner, who argues that if the diet, drug, or vaccine is going to benefit his animals, then all of them should have it. With field trials that involve people, the emotional factor is paramount and Sinclair Lewis, in his book *Martin Arrowsmith*, recorded very clearly the conflict between a research worker fighting for the truth, and his companions, moved by pity to take any step, however illogical, to reduce the suffering caused by an outbreak of plague.

In experimental work the " post hoc " argument is avoided wherever possible and it is usually replaced by the " method of difference." This method lays down that if you have two sets of circumstances that are alike in every respect except one, and if the phenomenon under investigation occurs in the group containing this one circumstance, then this circumstance is the cause, or the effect, or is part of the cause or effect of the phenomenon under investigation.

Assume that we were going to use 20 mice in a simple vaccination experiment; that all the mice were alike, and that they would be kept in the same environment. We would split the mice into two groups of 10 and give the vaccine to one group. Later we would give a test dose to all the mice. If the ten vaccinated mice survived and the ten unvaccinated, or control, mice died, then we could say that the single circumstance that was different—the injection of a vaccine—was the cause or effect, or part of the cause or effect, of the phenomenon under investigation, which in this instance was the survival of the mice. Further, we could say that as the vaccine was given before the phenomenon occurred it was the cause, and not the effect, of the phenomenon.

Statistical method

Unfortunately with biological material the " method of difference " is often an ideal that we cannot reach. The mice may all look alike but some, because of their physiological make-up, will be more susceptible or less susceptible than others, while in addition some may have suffered from mild and undetected diseases that have increased, or decreased, their natural susceptibility. Thus when a test dose is given, instead of having 10 deaths in one group and 10 survivors in the other, we may find that we have both dead mice and live ones in each group. It is with this type of result that the trained research worker uses a statistical test to provide an objective assessment of the result.

Because of these hidden factors we can regard mice as being like playing cards, with their outward appearance as similar to the backs of a pack of cards and their intrinsic values as varying as the unseen faces of the cards. Thus a card might belong to a black suit or a red one, it might bear a high number or a low one, and the number might be divisible by 3 or 5. Each of these possibilities might represent some crucial factor in the game to be played.

If, using cards, our experiment was to find which was the better of two players, we could ask them to play but our judgement would be tempered by the fact that the winner, instead of being the better player, might merely have held the better cards. All we could do to prevent this happening would be to make sure that the cards were well shuffled before they were dealt.

Statistical tests are based on the knowledge obtained from such actions as selecting playing cards, tossing coins or withdrawing different coloured balls from a bag. The results from such frivolous material have proved so useful that they are widely used in industry where false assumptions are punished by financial loss. A fundamental condition of this method is that before any card is selected the pack must be well shuffled; similarly the balls in the bag must be well mixed or the coin tossed in a fair manner.

If, therefore, we are going to regard the mice as similar to playing cards, and interpret the result by statistical methods, the theorist can maintain that the mice must be distributed into the two groups by some method by which they are shuffled, mixed or tossed; this process is known as randomisation. Methods of tossing or mixing mice might add excitement to laboratory life, but the same result is obtained by catching mice in sequence and deciding their group by tossing a coin or using some other method of chance.

With the contents of each group chosen by chance we apply our experimental interference to one group—the experimental group—and leave the other group—the control group—untreated. To judge the result we adopt the line of reason that, as circumstances vary in each group, we cannot use the method of difference. We can, however, say that we distributed the unknown circumstances as equally as possible into the two groups by using a method of chance. Therefore if, when the experiment is ended, one group varies from the other to a degree that would be unlikely to be due to chance alone, then there is an objective reason to accept that the treatment had produced some effect.

Thus the result has not been proved, if we had to do that the advance of science would be very slow, but the hypothesis has been justified, and Arber (1954) writing on the philosophy of science, accepts the ruling that all that is necessary for a forward step is that an hypothesis should be justified.

Selection into groups

To make the experimental and the control groups comparable there are two methods of selection. First, with measurable qualities we can try and keep the two groups as equal as possible and, second, where qualities are unseen and unmeasurable, we can distribute them by chance, making sure that each mouse has an equal chance of falling into the experimental or control group.

It is, of course, no good taking into account measurable qualities that experience has shown will not affect the result. For example, although all the mice may look alike, accurate weighing would show that each mouse was of a different weight, but unless we believed that this would affect the result, this variation would be ignored. The only measurable factor worth taking into account might be the effort made by each mouse to evade capture, and if it was felt that this alertness was an indication of good health, the following method of selection could be used. Prepare a cage for each of the two groups. Catch two mice and toss a coin to decide which cage the first mouse should be put into; putting the second mouse into the other one. Continue this sequence until all the mice are allocated and then decide by a final toss which lot should act as the experimental group.

Analysis

After allocating the 20 mice into experimental and control groups, the experimental mice will receive the vaccine and, after this has had time to become effective, the test dose will be given to all the mice. At an appropriate time after this the survivors in both groups will be counted.

Turning now to the analysis of the vaccination experiment, let us assume that in the result two vaccinated and six control mice have died. This type of result is often put into what is called a " two by two " or a four-cell table, as follows:

	Died	Survived	Sub-totals
Vaccinated	2	8	10
Controls	6	4	10
	8	12	20

A worker looking at these results might say " As I arranged these two groups to be comparable, my results show that, as I expected, the vaccine has given some protection." If asked about the effect of chance, and if at the stage where he still transferred everything to percentages, he might present his results thus " 100 per cent more mice have survived in the vaccinated group compared with the control group, or taking it the other way round, 200 per cent more mice have died in the control group than in the vaccinated group, which to me appears significant." With similar figures and a strong desire to be right, many long arguments can ensue if the worker is once put on the defensive.

To avoid these arguments, statisticians have laid down an objective standard as a guide as to whether an experiment has justified the hypothesis or not. This standard lays down that an experiment has not justified the hypothesis unless the figures in the result could not occur by chance alone more than once in twenty trials.

This criterion is based on a " Null Hypothesis " by which it is assumed that the experimental interference of vaccination has had no effect whatsoever. The deathrate in each group is therefore a matter of luck, dependent on how the naturally resistant mice were selected into each group. If, with 8 susceptible mice and 12 resistant mice, the number of survivors in each group could easily occur by chance alone then the null hypothesis is sustained, and there is no reason to alter the assumption that the vaccine has played no part in the result. If, however, the figures in the result could occur only once in 20 trials the result is accepted as sufficiently unusual to justify the assumption that some other factor has played a part. Therefore, if—and only if—the two groups have been treated in such a way that the only real difference in treatment between them was the vaccination of one group, then it is assumed that this vaccination has influenced the result.

Trial and error method

How can we find the likelihood of the various ways in which the 8 susceptible mice fall into the two groups of 10? If we obtain 20 rubber stoppers of the same size and mark 8 of them to represent the susceptible mice, we can use these symbols to repeat our experiment as many times as we like. The stoppers are placed in a bag which is well shaken to mix them up and the withdrawals are made blindfold. To imitate the first method we used with mice, one stopper would be withdrawn from the bag and a coin tossed to see if it was in the " heads " or " tails " group and then a second stopper would be withdrawn and placed in the other group. This procedure would be continued until all the stoppers had been dealt with, when the result would be recorded.

With a little experience it would be realised that this close imitation of the experimental method was unnecessary and that similar results could be obtained by picking out unseen stoppers several at a time, until 10 had been withdrawn. To build up what is called a " Frequency Table " we would record the number of marked stoppers out of the 10 chosen. First we would draw up a table with columns headed 1, 2, 3, . . . up to 8. We would then mix the stoppers in a box and withdraw 10 blindfold. We would then count the number of marked stoppers present, and if there were four we would make a mark under the 4 column. Finally, we would count the number of strokes under each column and these totals would give a frequency table. Using this method 100 trials were carried out with the following result:

Number of Marked Stoppers Withdrawn	0	1	2	3	4	5	6	7	8	Total Trials
Frequency with which they were withdrawn	0	0	9	30	28	26	7	0	0	100

Having obtained this frequency table we can, if we wish, fill in the other events. Thus the result of withdrawing no marked stoppers signifies that there were no deaths in the first group of 10 and that here there must be 10 survivors. As there were no deaths in the first group then the second group must contain all the 8 deaths and only 2 survivors. Working on the same lines for the other possibilities we can build up the following table:

Selected Group	Deaths	0	1	2	3	4	5	6	7	8
	Survivals	10	9	8	7	6	5	4	3	2
Remaining Group	Deaths	8	7	6	5	4	3	2	1	0
	Survivals	2	3	4	5	6	7	8	9	10
Frequency		0	0	9	30	28	26	7	0	0

From this table it can be seen that our result of 2 deaths in one group, which necessitates 6 falling in the other group, has occurred 9 times with the 2 falling in the selected group and 7 times with the 2 falling in the remaining group, giving a total of 16 times in a 100. Hence instead of occurring only once in 20 times, as set by the criterion, our result would occur about once in every 6 trials; thus our experimental trials have shown that such a result could happen fairly often by chance alone and that there was no need to postulate that the vaccine had produced any effect at all.

Permutations

We have discussed an experiment involving mice and have used the figures suggested for the result as the basis of another experiment in which, by using stoppers as symbols, and repeating the experiment 100 times, we were able to demonstrate that the result could not be considered statistically significant.

Must we always use stoppers, or is there an easier and more reliable arithmetical method of working out the odds ? Yes, there is an easier way which makes use of permutations, so that perhaps it would be best to remind readers what this term means before describing the test. Permutations are the numbers of different ways in which the number of objects you are dealing with can be arranged.

If the number of objects is 5, then you can have any of the 5 as the first object and any of the remaining 4 as the second object, so that each of the first 5 ways has 4 alternatives as second choice, giving 5×4, or 20, ways of arranging the first 2 objects. With each of these 20 ways there are 3 different objects possible for the third position and 2 alternatives for the fourth position with the remaining object for the fifth and last position. The possible ways of arranging them are; $5 \times 4 \times 3 \times 2 \times 1$, which is called " factorial five " and written either as $\underline{5|}$ or $5!$. Thus with the 5 letters in a bag the chance of spelling a word like " brain " correctly by pulling the 5 letters at random from a bag would be $\frac{1}{5!}$ or 1 in 120.

In some other five-letter words the same letter may occur more than once; for example the word " peels " has 2 letters the same. It seems obvious, and if it is not you can prove it by experiment, that there is the same chance of spelling " peels " correctly as there is of obtaining some other order of letters such as " plsee," and this latter combination is more convenient for explanation.

If we examine the chance of spelling " plsee " correctly, it can be seen that the chance of getting the *p* out of the bag first is 1 in 5, the chance of getting the *l* second, is 1 in 4 and the *s* third, is 1 in 3, but as the 2 *e*'s are the same, the rest of the factorial is not necessary. So that for " peels " (or plsee) we have a chance of $\frac{1}{5\times4\times3}$ but as you can look factorials up in tables to find their values, it is easier to write this $\frac{2!}{5!}$. In the same way, with the word " reeve " the chance of selecting the correct letters would be $\frac{3!}{5!}$, or 1 in 20.

As permutations and combinations are used for many simple problems of chance, and as they form the basis of other non-parametric tests besides the exact treatment, the research worker should have some idea of their use. Simple explanations are given by Levy and Preidel, Moroney, or Wesley, in the books mentioned in the bibliography.

Exact test

It has been found by experience that a formula involving the use of permutations can be used to solve problems of the 2×2 or four-cell type: Fisher (1938) calls this test the exact treatment. Before showing how this test can be applied to the mouse experiment let us see how it works out on a more simple example.

Imagine an experiment on 2 groups, each of 3 animals, which resulted in 2 animals being killed and 4 remaining alive. In our working model we could make this 2 marked stoppers which we can call *A*s, and 4 plain stoppers which we can call *B*s. If we tried 100 trials with these stoppers, and in each trial removed 3 and left the other 3 in the bag, the result of every trial would fit into a four-cell table in which the total and sub-totals were like this:

	A	*B*	Sub-totals
Withdrawn	*a*	*b*	3
Left in bag	*c*	*d*	3
	2	4	6

For these figures are fixed, either by our decision to have 3 in each group, or by fate in deciding that 2 animals would die.

With our 100 trials we would not only obtain every set of possibilities that these 4 empty cells can contain but we would also get some idea of their relative likelihood. It will be seen that the 3 sets of possibilities that can fit into these empty cells are as follows:

$$\begin{matrix} 0 & 3 \\ 2 & 1 \end{matrix} \qquad \begin{matrix} 1 & 2 \\ 1 & 2 \end{matrix} \qquad \text{and} \qquad \begin{matrix} 2 & 1 \\ 0 & 3 \end{matrix}$$

To work out the relative likelihood of these possibilities by the exact treatment instead of using a trial and error method, we follow the formula given by Fisher (1938). This formula is in two sections.

The first part of the formula is:

$$\frac{\text{Factorials of the sub-totals}}{\text{Factorial of total}} \quad \text{which in our example is:}$$

$$\frac{3!\,3!\,2!\,4!}{6!} \quad \text{or} \quad \frac{12}{5}$$

The complete formula is:

$$\frac{3!\,3!\,2!\,4!}{6!} \left\{ \frac{1}{0!\,3!\,2!\,1!} + \frac{1}{1!\,2!\,1!\,2!} + \frac{1}{2!\,1!\,0!\,3!} \right\}$$

Working this out in the form of a table we have:

Cells	Contents	Factorials	Products as Reciprocals	×	Result for Sub-totals		Chance	Chance as Decimal Fraction
a	0	1						
b	3	6	$\frac{1}{12}$	×	$\frac{12}{5}$	=	$\frac{1}{5}$	0·2
c	2	2						
d	1	1						
a	1	1						
b	2	2	$\frac{1}{4}$	×	$\frac{12}{5}$	=	$\frac{3}{5}$	0·6
c	1	1						
d	2	2						
a	2	2						
b	1	1	$\frac{1}{12}$	×	$\frac{12}{5}$	=	$\frac{1}{5}$	0·2
c	0	1						
d	3	6						
					Total		1	1·0

In this table we see that for the first set of possibilities the 4 cells contained figures 0, 3, 2, and 1. In the next column we see that the factorials of these, 0!, 3!, 2! and 1! amount to 1, 6, 2 and 1, the factorial of 0 being 1 (if you need proof of this look it up in a book on mathematics). In the next column the factorials are multiplied together, giving $1 \times 6 \times 2 \times 1$ or 12 permutations, giving a chance of $\frac{1}{12}$. This chance out of the $\frac{12}{5}$ chances, given by the sub-totals, equals $\frac{1}{5}$ so that $p = 0{\cdot}2$.

From this table we see that all the chances added together come to 1, or certainty. We also see that the figures 0, 3, 2, 1, come twice, once at one end with 0 and 3 in the withdrawn group, and once at the other end with 0 and 3 left in the bag. This means that if the result of our experiment was 0 and 3 in one group, and 2 and 1 in the other, the chance of our obtaining these figures would be $0{\cdot}2 + 0{\cdot}2$ or $p = 0{\cdot}4$.

Returning to the mouse experiment, it can be seen that this would involve some big factorials and that the number of multiplications would be very tedious. To avoid this we can use logarithms both for routine multiplication and for the values of the factorials. Tables giving these values are provided by Fisher and Yates in their statistical tables, and by Mainland (1948) in a very useful reprint that we will refer to again.

Before looking at the test it might be advisable to revise our knowledge of the value of logarithms to the base of 10. The logarithm of 2 is roughly 0·3. In the same way 20 is 1·3 and 200 is 2·3. Going in the other direction the logarithm of 0·2 is $\bar{1}{\cdot}3$, of 0·02 is $\bar{2}{\cdot}3$ and of 0·002 is $\bar{3}{\cdot}3$.

Proceeding with the test in the same order that we used before we see that the totals and sub-totals gave the following figures:

		10
		10
8	12	20

Using Mainland's table to find the logarithms of the factorials for these figures we add the logarithms of the sub-totals together in order to multiply them, we then divide this product by the factorial of 20 by subtracting its logarithm, so we have:

		Logs of factorials
	10!	6·5598
Sub-totals	10!	6·5598
	12!	8·6803
	8!	4·6055
	Product	26·4054
Total	20!	18·3861
		8·0193

We now have this logarithm of 8·0193 to use in the same way as we used the fraction of $\frac{12}{5}$ in the simple example, and we use it to multiply the chance given by each set of possibilities. Recording this as a table we will have the figures given in Table 15.

It can be seen that in Table 15 the whole range of possibilities is worked out simply by adding or subtracting the logarithms found in a table, and that when all these possibilities are added together they should amount to one. In Table 15, due to the error present in four-figure logarithms, the total has come to 1·0001; the smallness of this error checks the arithmetic. It can be seen that the results repeat themselves after reaching the middle so that only half the series needs to be calculated.

Double-sided significance

It has already been stated that in order to have some objective criterion, a chance of 1 in 20 has been chosen to signify significance; similarly a chance of only 1 in 100 is usually accepted as highly significant; what we are referring to here is double-sided significance.

To get some idea of how this is measured one can use the analogy of firing at a target that consists of a bull's eye surrounded by concentric circles which delimit the score. If we tested 400 rifles by fixing them into the same position and firing one shot at the target, we might find that 4 shots were outside these circles. It could be that one of these shots was outside at the top, another outside at the bottom, another on the left, and the last one on

TABLE 15

In Logarithms

Sets of Possibilities		Product of Factorials	$\frac{1}{\text{Product}} \times \frac{\text{Sub-tot.!}}{\text{Total!}}$	Antilog. or Chance
0		0·0000		
8		4·6055		
10		6·5598	8·0193	
2		0.3010	−11·4664	
		11·4664		
			$\bar{4}$·5529	·0004
1		0·0000		
7		3·7024		
9		5·5598	8·0193	
3		0·7782	−10·0404	
		10·0404		
			$\bar{3}$·9789	·0095
2		0·3010		
6		2·8573		
8		4·6055	8·0193	
4		1·3802	−9·1440	
		9·1440		
			$\bar{2}$·8753	·0750
3		0·7782		
5		2·0792		
7		3·7024	8·0193	
5		2·0792	−8·6390	
		8·6390		
			$\bar{1}$·3803	·2401
4		1·3802		
4		1·3802		
6		2·8573	8·0193	
6		2·8573	−8·4750	
		8.4750		
			$\bar{1}$·5443	·3501
5				
3			$\bar{1}$·3803	·2401
5				
7	6			
	2		$\bar{2}$·8753	·0750
	4			
7	8			
1				
3			$\bar{3}$·9789	·0095
9	8			
	0			
	2		$\bar{4}$·5529	·0004
	10			
				1·0001

the right. We could look at this result from two viewpoints. Firstly, we could say that rifle number, say, 107 was very bad because out of a sample of 400 it shot most to the left, and that number 302 was very bad because it fired higher than the other 399, and so on with the other 2 rifles. But, secondly, we could take a more practical point of view and ignore the precise direction in which they were inaccurate, and refer instead to 4 rifles out of 400 that fired shots more than 12 inches away from the bull's eye. This second method is the one used in statistics, but, as usually there can only be figures that are too high and figures that are too low, the concentric circles can only cut off two extremes, so that this is spoken of as double-sided significance. If, for some special purpose, we wish to refer to an item as being the largest (or the smallest) item in a hundred, then we are using single-sided significance, for here we ignore the complementary item at the other extreme. We need to know this because some tables may refer to single-sided significance instead of the conventional double-sided significance.

Result of the exact test

For convenience the results obtained in Table 15 are summarised below together with the result of the trial and error method.

Deaths in One Group	Theoretical Frequencies from Exact Treatment	Double-Sided Probability	Actual Frequencies with Trial and Error
0	0·0004	0·0008	0
1	0·0095	0·0198	0
2	0·0750	0·1698	9
3	0·2401	0·6500	30
4	0·3501		28
5	0·2401	0·6500	26
6	0·0750	0·1698	7
7	0·0095	0·0198	0
8	0·0004	0·0008	0
	1·0001		100

Working on the analogy of a target we can place the outermost circle so that it excludes only the end results of none dead in one group and 8 dead in the other; it will be seen that we exclude

0·0004 at each end so that $p = 0{\cdot}0008$. This is well under 1 in 100 and this result would be highly significant. Our second circle would also exclude 1 dead in one group and 7 in the other, and here the probability would be twice 0·0095+0·0004 giving $p = 0{\cdot}0198$. This is greater than $p = 0{\cdot}01$ so that the result is not highly significant, but it is well under $p = 0{\cdot}05$ and therefore the result is significant. The next circle would exclude 2 in one group and 6 in the other; this would add the chance of 0·0750 at each end, so that the total would be 0·1500+0·0198 giving $p = 0{\cdot}1698$. This is roughly 17 in 100 and, as this is more likely to occur than 1 in 20, the result is not significant. Note that 2 dead in one group and 6 in the other was the result for the mouse experiment, so that it can be seen how well the theoretical value of 17 in 100 agrees with our trial and error answer, which was 9 one side of the mean and 7 the other side, or 16 in 100.

Learning mathematics

Lancelot Hogben in *Mathematics for the Million* likens mathematics to a language. This is so and one similarity is that you cannot learn either just by reading a book about it. To learn either you must make yourself familiar with the verbs or operators and the nouns or numbers. To read the Latin for "The daughters of the queen wash the feet of the sailors with water" may convey a beautiful concept but it rather restricts conversation unless you know how to transpose it to other situations, such as "The daughters of the queen, by the sailors . . ." and so on. Similarly, when you read the mathematical parts of this book you should have a pencil and paper near you so that you can try out the operations with other simple figures. Thus when learning the possessive, stick to "The pen of my aunt" and do not embarrass the memory with "The writing instrument of my maternal parent's female sibling." To get the idea of $\frac{1}{2}$ of $\frac{1}{3}$ is fairly easy, to get the idea of $\frac{1}{4{\cdot}935}$ of $\frac{1}{2{\cdot}879}$ may be frightening.

This is not primarily a book on applied statistics and exercises are not provided as it is not intended that the reader should confine his knowledge to this book alone. Books that may help the reader are given at the back of this book as a bibliography and this includes two very inexpensive books (Levy and Priedel, and Moroney) which provide both exercises and answers.

Tables of exact test and summary

It was shown how a simple small experiment may be planned and carried out using random selection. In order to keep our own wishes under control it was shown how the result could be judged by statistical methods. The first method was one of trial and error which could be carried out with a calabash and some pebbles even on a desert island, the second method was not too tedious provided tables of factorials were available. An even easier method can now be described, that is, to look up the result in a book of tables drawn up by Mainland, Herrera and Sutcliffe. The answers to experiments of this type involving equal numbers in both groups, up to 20 in a group, is given in their Table III and part of this table is recorded in the Appendix.

In this table N is the number of animals in each group, which in our experiment was 10. The number of A's in sample (1) over number of A's in sample (2) refers in our case to 2 dead in the vaccinated group and 6 dead in the controls. It then reads " Probabilities from one tail are in parentheses." This means we are dealing with single-sided significance.

Looking down the table we see in the third column $\frac{2}{6}$ (·0849).

If we double the number in this bracket to give the conventional double-sided probability we get $p = 0{\cdot}1698$, which is identical with our own result. The result of this experiment was therefore not significant, for to be significant the result should only occur once in 20 times ($p = 0{\cdot}05$) by chance alone, and to be highly significant, only once in 100 times ($p = 0{\cdot}01$).

Other tables in the book include selected probabilities for unequal numbers up to 20 in one group.

References

Fisher, R. A. (1938). *Statistical Methods for Research Workers.* Oliver and Boyd.

Mainland, D. (1948). *Statistical Methods in Medical Research.* Canad. J. Res., E, 26, 1-166.

Trotter, W. (1916). *Instincts of the Herd in Peace and War.* Fisher Unwin.

Chapter 2

A FIELD EXPERIMENT AND THE CHI-SQUARED TEST

The practical man is one who practises the errors of his forefathers.

T. H. Huxley

In the first chapter we spoke of an experiment in which the result depended on the number of mice that fell into one of two categories the quick or the dead. This type of data is called qualitative, or enumeration data, and the items are not measured but are classified on the " passed " or " failed " basis. The little Victorian servant girl, who confided to her mistress that she was just a tiny bit pregnant, would fall straight into the " pregnant " class. Another, and a better known, test used for this type of data is the chi-squared test (ch as in chemist) and the application of this test will be described.

Later in the book experiments will be described step by step, but in this chapter we will have a broad look at a field experiment because many biologists are called upon to make such experiments, and because they expose many practical details that are present, but more hidden, in laboratory experiments.

Many horrible things are called " field experiments," for they are the playground of the " practical man," and the results are rarely modest. For example, here, disguised but not exaggerated, is a result from an article published in a recent journal. " Although the clinical measurements are not definite enough for us to be dogmatic, and no controls were used, we feel that we can make the following recommendations." One sees at once that the worker has imagination and self-confidence, but one loses one's own confidence in both him and the journal. It is like a civil engineer building a bridge and then saying that he doesn't know how strong it is but that he would recommend it for fast traffic.

As the basis for our experiment let us assume that we are dealing with a disease of cows that may cause them to abort at about the sixth month of pregnancy, as compared to the normal pregnancy of over nine months, and that we have been asked to test the effect of a new drug called " Stayput."

When in charge of an experiment of this type, one's first impulse may be to visit the herd-owner as quickly as possible in order to make personal contact and perhaps to start dosing the cows before any further abortions occur. It must be remembered, however, that a herd-owner can only remain a herd-owner by overcoming a long series of practical difficulties and in doing so he has become resistant to sales talk and has not learned to suffer fools gladly. Hence, before going to see him it is wise to find out what he has already been told by your senior worker and also to brush up your knowledge of both the disease and the remedy to be tested, for when you do arrive you will be subjected to a series of questions, such as: " What is this new drug ? ", " Will it lower the milk yield ? ", " Will it taint the milk ? ", " Who else near here has got this disease ? ", " What is he doing about it ? ". " How did my cows get this disease ? ", " Is it true that feeding potatoes and salt prevents this disease ? ", and so on. An enthusiastic visionary can do a lot of damage to the farm routine and the milk yield, and a cross-examination to discover the worker's knowledge of his subject is the herd-owner's first line of defence. Hence, before you visit the herd, know your subject, know what you want to do, know what information you are going to ask for and have the appropriate columns ruled in a book to make the questions obvious and the answers easily recorded.

The problem of a control group

The method of reasoning to be employed in this experiment will be a rather battered form of the " method of difference " and therefore, after a general conversation with a herd-owner, the first thing to find out is his reaction to the suggestion that you wish to treat half the cows only and leave the other half as controls. If approached directly on this question some owners will react immediately and make it obvious that under no circumstances will they permit this. The layman usually has the impression that although some treatments are better than others, there is always a remedy for every disease, and if you do not know what it is then he will have to find out from somebody else. The idea that you are only going to apply your treatment to half his herd and then callously stand by and watch the other half suffer, just so that you can make a measurement, is more than he can stomach. The fact that the advertised remedies for this disease are all carefully worded with such phrases as " the impression gained by Dr. Hope " or " From field trials we believe . . . ", indicating as they do, that the results were probably due to chance alone, mean nothing to him, and if you refuse to treat all his cows he will go elsewhere for a treatment that can be given to the whole herd.

Unfortunately this type of herd-owner is not exceptional and this difficulty is one of the biggest problems in field trials. A further point is that even if you find an owner who will agree to leave half the cows as controls, this may not be without complications, for when the owner goes round with you when you are randomising the cows, he may break in every now and again to demand that this particular cow must receive treatment, until all his best cows have found their way to the treated group. Further, as the experiment will probably last over a year, each month may show that one or two of the control cows have been sold so that the control group becomes so small that it is unreliable to predict from it. Finally, if the cowman can lay his hands on the remedy, then, if he is a good cowman, he will give the remedy to any cow in the control group that looks sick, even if it means a healthy looking experimental cow going without.

It is obvious from what has been said that, even if the owner agrees to the treatment of only half the herd, factors may be introduced which damage the accuracy of the result, but no error is as great as that of complying with the owner's request to treat all the cows the same way. Wishful thinking may suggest that it will be possible to compare the result with that obtained in an untreated herd a few miles away; you may wish to believe that, if abortions stop suddenly, this was due to your treatment; or you may like to believe that you can compare your results with an accepted abortion rate. None of these methods is acceptable for the result cannot be measured. Bradford Hill (1946) writes of biometrical tests being " arithmetic with logic " and if you begin by knocking the logic out of your experiment the most advanced mathematician cannot help you, and you must go for advice to an astrologer. If you believe that perhaps, if you treated two or three herds, your evidence would become sufficiently accurate, then you may be interested in the verdict of Topley and Wilson (1941) who weighed the evidence on the immunity given to children in an experiment using B.C.G. vaccine, in which the environment of the vaccinated children was not strictly comparable with that of the controls. They said: " For these reasons, the reproduction, even if it were possible, of the immense pile of data that has accumulated on this subject in France and French-speaking countries, would serve no useful purpose, since from the statistical point of view it is practically worthless."

If at the outset, therefore, you sense that the dairy farmer will not allow only half his cows to be treated, it is best to avoid any argument and to go straight out for an experiment in which you will compare two different remedies. One half the herd will be treated with " Stayput " and the other with some other treatment. The control treatment might be the best advertised treatment then in vogue; it might be some form of antibiotic that the dairyman uses for every disease he meets; or it may be something that the owner has shown an interest in. Thus, if one of his questions was as to the value of salt and potatoes, then such a combination could be the basis of the control treatment. The use of two treatments will, of course, change the result of a successful experiment from the statement that the use of " Stayput " is justified, to one that " Stayput " is significantly better than the control treatment. How useless the control treatment is will be a matter for the conscience and ingenuity of the operator, but it is so difficult to find a really effective remedy for any disease that by and large a remedy will not be embarrassed by the competition of any control treatment. Once a control treatment has been chosen it must never be spoken of contemptuously either by the experimenter or his assistants.

After finding that some type of controlled experiment is possible the next step is to try to obtain some accurate information on the cows. The ideal would be to examine each cow and, after noting its number, add such information as its age, the time of service, its previous breeding history, its health history and any other information such as its breed, or whether it was bred in the herd or was bought in from somewhere else. This information recorded in appropriate columns, permits a more logical experiment to be planned and may also provide an interpretation of the result, or suggest another hypothesis.

Numbering of animals

Unfortunately people resent being asked for accurate information, perhaps owing to painful interviews with income tax inspectors, and the field worker may find it tactful to restrict his questions to the number of the cow, its age and probable date of calving. The quest for a number for each cow, must however, be relentless. You may well be told that each cow has a name plate standing directly in front of it, but it would be a very exceptional herd in which this was absolutely true, and sooner or later you would find a cow that had just been moved in, or where it had been changed for another, or you would be told that although the plate read "Daffodil II" the cow was really young "Daffodil III." Alternatively you might be told that you need not worry the slightest about numbers as Old Bob not only knows them all by name but knows the names of their dams and sires as well; in this case be sure that on some future occasion Old Bob will be away with lumbago or off to a wedding or a funeral, perhaps his own.

It is essential that if you are going to take any pride in the experiment the animals be numbered. If they are not numbered then permission should be asked to have them tattooed in the ear, or if the owner objects to a permanent number, to brand them on the hoof or the horn, and this applies to both experimental and control cows.

Administration of remedy

The responsibility of the administration of the remedy will depend to a large extent on how often it must be given. If it is to be given each day then perhaps it will be given in the food, and here it will be the responsibility of the field worker to mix the remedy with the diet and to see that a sufficient supply of the mixture is always available. In this instance the actual feeding of the remedy will be done by the cowman, so that some simple division of control and experimental animals must be arranged and complete randomisation, which would involve feeding one cow here and another one there, would be too exacting a task to ask him to perform each day. Here, the position of each cow, and its calving date, should be considered to see whether, by the movement of one or two cows, the remedial diet could be fed on one side, or at one end, of the byre and the other left as the control. It can be seen that to arrange such an experiment will take both ingenuity and salesmanship, for the movement of each cow will involve extra work until it has learnt its new position.

If the remedy is to be administered once a week then a competent field worker will either do it himself, or do it once a fortnight and get his assistant to do it on the alternate weeks. In this instance cows can be properly randomised and any appropriate restraints can be applied. It will be on these weekly or fortnightly visits that the recordings will be made. If the remedy is to be given less often than once a fortnight then the field worker should definitely be there at each administration.

Recording

In most experiments it is the measurements that are additional to the essential ones, that often turn out to be important, and it is these that often produce the paradoxes which lead to further fruitful investigation. In the experiment outlined in which the disease is characterised by the occurrence of abortion at the sixth month of pregnancy, it will probably be well over a year before all the cows and heifers have been tested and such additional measurements must be few for there is a limit to the inconvenience that can be placed on the herd-owner. Nevertheless, in addition to the recordings of calvings and abortions, it might be possible to record the milk yields, if these are kept, to note abnormalities connected with pregnancy and to carry out a post-mortem examination on dead calves.

It has been suggested that the experiment would last well over a year, which is long enough for both the field worker and the herd-owner to get heartily sick of it. It has been suggested that during this time the field worker should visit the farm regularly once a fortnight and while there he should not only obtain the records but that he should take some active part in the administration of the remedy, if only to accompany the cowman as he doles out the remedial and control rations, for it is during the casual conversation that accompanies such a task that he can learn from the cowman that vital information that is never obtained by direct questions; it may well be that in an experiment in which milk yields are important that on one occasion he will be told quite casually that an experimental cow is sucking the milk from the control cow that stands next to it.

The tendency is to visit a farm often when the work is new and interesting and then, as the work loses its interest, to cut down the number of visits. This must be avoided for it is the duty of the field worker to keep the herd-owner and the cowman interested, and if a dairyman finds that a field worker, who visited him regularly for the first three or four weeks, begins sending an assistant and makes his own visits less and less frequent, that cowman cannot be expected to retain any enthusiasm, and will consequently take all those short cuts that may ease his work yet jeopardise the result of the experiment. A field worker must, therefore, make up his mind that he is going to visit the herd regularly as a matter of discipline, and at the very beginning of the experiment should avoid setting himself a task that he is not going to be able to carry out.

From what has been written it can be realised that if you are going to carry out a sound field experiment you are going to do a lot of hard work. If it is not worth while making sure that the result is sound then why bother to do the experiment at all? To get the best result you will need experience, enthusiasm, salesmanship and discipline. Yet another factor in running a sound experiment is an expense account which can provide some small remuneration for those cowmen whose interest is vital.

Probably enough has been said to indicate that field experiments, although always a necessary step before the release of a remedy to the general public, are not the ideal form of research experiment and they cannot replace the properly controlled experiment in an experimental herd under the direct control of a research worker. Thus, although a purely experimental herd is a very expensive instrument, the expense of a poor field experiment can be a complete waste of money and may lead to such an ambiguous result that further investigation on the same problem is made necessary.

Let us imagine that the weary months have passed, that the

remedy has been given regularly, that we have managed to keep a distinction between treated and control groups and to record the incidents that have occurred. It is now the time to inspect, and where necessary, analyse the results.

The experiment may have been irksome but at least it got the worker out into the fresh air. Inspection and analysis of the records will give him no such relief and it represents a period of concentration, pencil biting and chain-smoking, that most workers would gladly avoid. In spite of this it is an essential part of the experiment, and if it is not done efficiently it will give your critics the chance to confute you with your own figures. In carrying out this analysis it is usual to begin by examining the main result and then to carry on by splitting the main result into parts, and, finally, to go on to the examination of those subsiduary results offered by the records.

Analysis of main result

Let us assume that during the period of over a year, 93 cows and heifers have calved with a high abortion rate, to give the following table:

	Abortions	Normal Pregnancies	
Remedy	8	38	46
Controls	14	33	47
	22	71	93

It can be seen that the results again provide a four-cell table. Unfortunately, as there are over 20 cows in each group, it will not be possible to look the result up in Mainland's table. The exact test is still applicable but would be tedious to apply, for even if one felt confident of one's arithmetic and neglected the check that the sum of all the probabilities came to one, it would still be necessary to work out the series of possibilities in which the smallest cell contained 8, 7, 6 . . . down to 0, giving 9 sets of calculations.

To avoid this tedium we propose to use the chi-squared test which was introduced by Karl Pearson and is based on the binomial distribution.

Binomial distribution

In the previous chapter we used the exact treatment, which can be described as a " direct method " because it can be worked directly from the combinations that are possible when drawing stoppers from a bag, in exactly the same way as the mice could have been selected at random in the experiment. In this method the selected balls are not replaced before the succeeding ball is selected.

In contrast the χ^2 test uses fixed odds, such as those given by tossing a coin or throwing a dice, or drawing balls from a bag, replacing each before the next selection was made. And the probabilities can be worked out mathematically using the binomial theorem. To compare the probabilities given by the two different methods we can first use the results of the mouse experiment where 8 mice died out of 20. With the exact test the chance of drawing 3 of the 8 marked stoppers from the bag, one after the other, would be $\frac{8}{20} \times \frac{7}{19} \times \frac{6}{18}$ or $p = 0{\cdot}049$. In contrast, with the binomial method, after each stopper had been drawn and the result recorded, it would be returned to the bag, and the contents remixed. As the odds remain the same the chance would be $\frac{8}{20} \times \frac{8}{20} \times \frac{8}{20}$, or $p = 0{\cdot}064$. Thus one method gives 5 and the other 6 in 100. With larger numbers however, the difference between the two methods is less marked. Thus, using the results for the field experiment, in which we assumed that there were 22 aborting cows out of 93, the chance would be $\frac{22}{93} \times \frac{21}{92} \times \frac{20}{91}$, or $p = 0{\cdot}012$ by the exact test, or $\frac{22}{93} \times \frac{22}{93} \times \frac{22}{93}$ by the binomial method, giving a probability of 0·013. So that with these larger figures one method gives 12 in 1000 and the other 13 in 1000, a very close approximation.

Pearson-Galton wedges

The binomial distribution is produced by the expansion of such expressions as $(p+q)^n$; should you wish to know more about this it is described in some of the books mentioned in the bibliography (e.g. Levy and Preidel; Wesley; etc.) but here it will merely be introduced in connection with a device used by Karl Pearson and Francis Galton to demonstrate the binomial distribution practically.

The model consists of a series of wedges surmounted by a funnel into which shot can be poured. Each wedge is placed in a position in which it will send one part of the shot dropped onto it to the left, and the other part to the right. If we assume that the wedges are placed to send just half the shot to the left and the other half to the right, then we have here the same proportions that we would obtain by tossing a penny, and the model would distribute the shot as shown in the following figure.

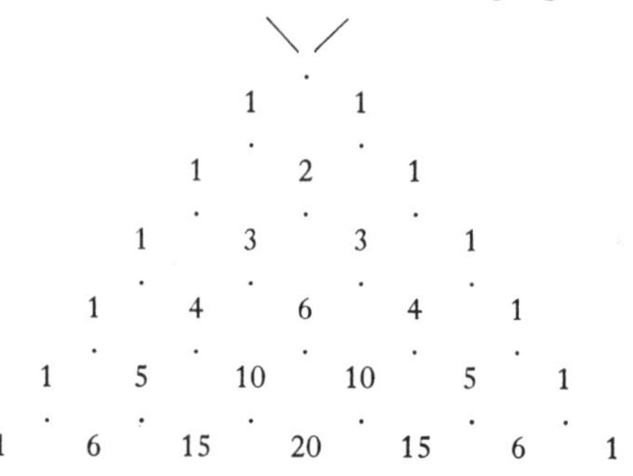

With this model half the shot will go to the left of the first wedge, and the other half will go to the right. This is equivalent to tossing one coin at a time, for this on the average would give the proportion of one head and one tail, which we could represent as $(H+T)^1$. Out of every 4 balls running down the funnel 2 will go to the left, and 2 to the right, of the top wedge. Of the 2 that go to the left, and hit the left hand wedge, one will go further to the left, and the other will be deflected back to the centre. This is equivalent to tossing 2 coins at a time, where on the average we would expect to get both heads once, both tails once, and 1 head and 1 tail twice, out of every 4 throws, in the same way that $(H+T)^2$ gives $H^2+2HT+T^2$. After passing the third line of wedges the results would be equivalent to tossing 3 coins at a time, and here, out of every 8 tosses, you would expect on the average to get all 3 heads once, and all 3 tails once, with the other results, half 2 heads with 1 tail and half 2 tails with 1 head. This would be equivalent to $(H+T)^3$ which when expanded gives $H^3+3HHT+3HTT+T^3$.

The figure ends with the results for 6 wedges, equivalent to tossing 6 coins at a time, and you can see that, because the work of 1 wedge is alternately counteracting or augmenting the work of another, we end with a definite pattern in which many items are near the middle, and the remainder thin out, step by step at the extremes. This is the binomial distribution.

This binomial distribution could be used to measure significance in the following way. The bottom line could refer to the results for tossing 6 coins at a time, and it is based on a total of 64 trials. The result for 6 heads turning up together would be represented by the 1 at one end or the other. Using double-sided significance we would say that an event as extreme as this only turns up twice in every 64 trials, that is, once in 32 trials, and as this is less often than 1 in 20 it can be regarded as significant.

Introduction to the chi-squared test

With a great deal of work it would be possible to build up binomial distributions line after line, until one could measure results for hundreds of coins thrown at the same time. But even this would not be enough for practical work, for they would only give answers for problems where the chances for and against an event occurring were equal. So that it would be necessary to start again, this time imagining, for example, that the wedges would throw 1 shot to the left for every 2 shots thrown to the right. It would only be when a table based on wedges that threw 22 shots to the left for every 71 thrown to the right, had been prepared, that we would have a table that would solve our field experiment in which 22 out of 93 cows aborted.

The making of these huge tables was made unnecessary by the invention of the χ^2 test, by which a close approximation of the probabilities of events distributed in a binomial manner, can be calculated from a simple formula. The basic formula for this test is $\frac{(A-T)^2}{T}$ where A is the actual number in the result to be be tested, and T is the theoretical number, either known or calculated from the data.

To demonstrate the test let us go back to the figures shown in the table given by the Pearson-Galton wedges. It will be seen that in this the total number of trials in each line is double that of the previous line, and that the end figures are always one, and that the figure next to them is always the same as the serial number of the row. If, therefore, we were to use 12 coins at a time, the number of trials in the total would be 64 for the 6 coin line, multiplied by 2 six times, to bring it up to the 12 coin line; this gives 64×2^6 or 4096. Out of 4096 trials as many as 11 heads (or 11 tails) would occur $1+12$ times, the same distance from the average would exclude 13 at the other end as well. This would give 26 out of 4096. This amounts to 0·635 per cent. or roughly 6 in 1000.

Using the χ^2 test this probability would be estimated as follows. The result of 1 head and 11 tails would be compared with the expected result of half and half as the theoretical, using the formula $\frac{(A-T)^2}{T}$ for each part of the result. This would give:

$$\chi^2 = \frac{(1-6)^2}{6} + \frac{(11-6)^2}{6} = \frac{50}{6} = 8{\cdot}33$$

If we then look up this value in Table 4B given by Yule and Kendall, we find that where $\chi^2 = 8{\cdot}3$, $p = 0{\cdot}00396$. This is roughly 4 in 1000. Thus by a short test, taking a minute or two we have with the help of a table, found a probability of 4, which is a very good approximation of the first result of 6 in 1000.

Yates' modification

With large numbers the agreement between χ^2 probabilities and those given by the binomial distribution is very good, but with small numbers, where for example the theoretical number is as low as 6, the χ^2 test sometimes gives a higher significance than that given by the distribution; and this is a fault on the wrong side.

Using 12 coins we found good agreement when we assumed a result of 1 head and 11 tails, but if we had gone on to ask: How often is there a result as extreme as 2 heads and 10 tails? Then the χ^2 test would have shown a higher significance. Thus the extreme figures were 1 and 12 for no heads or 1 head, if we add the next figure for 2 heads we find that this is 66, giving 79 at each end, or 158 in 4096. This comes to 3·86 per 100, or $p = 0{\cdot}0386$.

With the χ^2 test we have $\frac{(2-6)^2}{6} + \frac{(10-6)^2}{6} = \frac{32}{6} = 5{\cdot}33$. Yule and Kendall's table gives $\chi^2 = 5{\cdot}3$ for $p = 0{\cdot}02133$, a difference of nearly 2 in 100 on the side that overestimates significance.

To overcome this error, found when using small numbers, Yates introduced a modification by which each result was taken half a unit nearer the average. Using this modification we would test the 2 heads and 10 tails as if the result had been $2\frac{1}{2}$ heads and $9\frac{1}{2}$ tails. This would give:

$$\chi^2 = \frac{(2\frac{1}{2}-6)^2}{6} + \frac{(9\frac{1}{2}-6)^2}{6} = \frac{2(12{\cdot}25)}{6} = \frac{24{\cdot}5}{6} \text{ or } 4{\cdot}1.$$

From the table $\chi^2 = 4{\cdot}1$ where $p = 0{\cdot}04288$. Thus both this result and the one from the distribution approximate to 4 in 100 and the error is reduced to less than 5 in 1000.

It is always advisable to use Yates' modification, for if the figures are critically few, it enables one to predict with caution, while if the figures are ample, it makes very little difference to the result.

Chi-squared test

We now return to the main result of the field experiment, which we said gave the following four-cell table.

	Ab.	*N.P.*	
Treated	8	38	46
Controls	14	33	47
	22	71	93

We tried to have equal numbers in each group, and ended the experiment with 46 and 47. Time supplied the answer that 22 cows aborted and 71 cows did not. These are the facts that supply the sub-totals, and from these we must try to calculate how likely it would be for the cell contents to be as they are if the treatment had had no effect whatever.

The null hypothesis, therefore, is that these results are equivalent to those obtained by pouring one group of 46 shot, and another group of 47 shot, over a single wedge designed to throw, on the average, 22 shot to the left and 71 shot to the right. If we used this device, would the contents of the cells be the sort of numbers we would expect?

Before we can calculate the value of χ^2 we must find the theoretical values for the four cells. If the ratio of abortions is 22 out of 93, then the top left hand cell will have a theoretical value of $\frac{22}{93} \times 46$ so that T will be 10·9. If we know the top left hand cell contains a T value of 10·9 then, because we also know the four sub-totals, we can find all the other T values by subtraction; thus to make a sub-total of 22, the bottom left hand cell must have a value of 22—10·9, or 11·1.

Having found the theoretical values we can next apply Yates' modification, which would give the following figures:

$$\begin{matrix} 8\frac{1}{2} & 37\frac{1}{2} \\ 13\frac{1}{2} & 33\frac{1}{2} \end{matrix}$$

We now apply the formula of $\chi^2 = \frac{(A-T)^2}{T}$ to each cell:

Modified Cell Content	Theoretical Value For Cell	$\frac{(A-T)^2}{T}$	X^2
8·5	10·9	$\frac{(-2{\cdot}4)^2}{10{\cdot}9}$	0·53
13·5	11·1	$\frac{(-2{\cdot}4)^2}{11{\cdot}1}$	0·52
37·5	35·1	$\frac{(-2{\cdot}4)^2}{35{\cdot}1}$	0·16
33·5	35·9	$\frac{(-2{\cdot}4)^2}{35{\cdot}9}$	0·16
			1·37

The formula is easily worked out with a slide rule. The cursor is brought up to 2·4 on the bottom scale D, which gives $(2{\cdot}4)^2$ on the top scale (A). Then 10·9, 11·1, 35·1, and 35·9 on the upper scale of the moving part (scale B) are brought up to the cursor in turn, and the reading on scale A standing above the 1 in scale B is recorded for each.

Quick formula

Mathematicians dislike doing unnecessary work and are prepared to puzzle out formulae that reduce labour, even if these obscure the basis of the calculation. A quick formula for the four-cell test is given below, and is demonstrated on the data used in the previous example. It can be seen that the formula incorporates Yates' modification in its make up so that when using it there is no need to adjust the cell contents individually.

	Ab.	N.P.	
Treated	8 (a)	38 (b)	46 ($a+b$)
Controls	14 (c)	33 (d)	47 ($c+d$)
	22 ($a+c$)	71 ($b+d$)	93 (T)

Using the bracketed symbols seen in the Table this formula is:

$$\chi^2 = \frac{((ad \sim bc) - \frac{1}{2}T)^2 \times T}{(a+b)\,(c+d)\,(a+c)\,(b+d)}$$

$$= \frac{((8 \times 33 \sim 38 \times 14) - 46{\cdot}5)^2 \times 93}{22 \times 71 \times 46 \times 47}$$

$$= \frac{(268 - 46{\cdot}5)^2 \times 93}{3377044}$$

$$= 1{\cdot}35$$

Putting this in words, we multiply the two diagonals and subtract the smaller product from the larger one. We then carry out Yates' modification by subtracting half the total, square the result and multiply by the total. We then divide this by the product of all the sub-totals multiplied by each other. Often it is possible to cancel out some of the items. Using a slide-rule it would be best to leave the numerator as $221{\cdot}5 \times 221{\cdot}5 \times 93$, and then alternately multiply and divide along the fraction, using the sub-totals as they stand.

Statistical tables

By using the quick formula we obtained an answer of $\chi^2 = 1 \cdot 34$, a little different from that of the step by step method, where the *T* values were rather approximate. We now need a table that will show us the probability denoted by this value. So far we have been looking up χ^2 values in Yule and Kendall's table, but this was because we wanted the probability of a double figure number, but this accuracy is seldom required. Statistical tests are of different types, and what we will want is a set of statistical tables which give selected figures for judging as many different tests as possible. Such a set is the book of statistical tables produced by Fisher and Yates, which we mentioned already as containing factorials for the exact test, and we will refer to these tables from time to time as we deal with other tests.

In their book the table for χ^2 gives a series of selected probabilities along the top, including $p = 0 \cdot 05$ for significance, and $p = 0 \cdot 01$ for highly significant, and down the side are different values of n, which represent degrees of freedom. Part of this table is seen in an abridged form in the appendix of this book.

Degrees of freedom can be likened to the number of comparisons that can be made. In the field experiment the number of items that fell on, say, the left side of the wedge were compared with the number that did so in the second group. This represents one degree of freedom. It is the simplest comparison possible, for if we remove either group, or the wedge, there is nothing left to compare. On the other hand, if we add another group, or an extra wedge, then we will be able to make an extra comparison and so we have obtained a second degree of freedom.

Looking at the χ^2 table for the position of our result of 1·34, we see that in the top row, which refers to one degree of freedom, the figures 1·074 for $p = 0 \cdot 30$, and 1·642 for $p = 0 \cdot 20$. Our result falls between these two, so that we can say that results similar to these can occur by chance alone once in every 3 to 5 trials. If we had used Yule and Kendall's table we would see that our result was about equal to once-in-four times, but this extra accuracy is of little or no value in an experiment of this type.

With the exact test we found that we could look up the actual result in the tables produced by Mainland, Herrera and Sutcliffe.

These tables also include some for the χ^2 test where the groups are of equal sizes, and where there is one degree of freedom. Using these tables we would take the abortion rates of 8 and 14 as representing *A*'s. Table 1, the 5 per cent table, does not give results for groups of 46 or 47, but those for 40 and 50 are recorded. With both 40 and 50 items in each group a ratio of 8/18 *A*'s is demanded for significance, so that we can see without calculation that our result is not significant.

Result of the experiment

An experiment has been carried out and the result was not significant, in fact calculation showed that it could occur once in 4 times by chance alone. Statistical tests are tools to help us think, not to stop us thinking, and it is up to us to use this information as we think fit. The fact that the result is not nearly significant need not prevent us writing a short article to record the result, for this, with tests by other individuals elsewhere, will help people to form an opinion. Alternatively we can use the figures obtained from this result to estimate the number of animals required to give a significant result, provided the same bias is present. Or we can alter the dose and try again. The one thing we must not do is to tell other people that the beneficial result was due to " Stayput," for we have seen that it could easily occur by chance alone.

Anything else of value that can be got out of the experiment will depend on the other recordings that were made, for example, the milk yields, or calving records.

Correct categories

The art of using statistics is mainly in the logic and not the arithmetic. For example we have used two tests and in both instances it was possible to test for significance by looking the actual results up in a table, and without making a calculation at all. Part of the art is to include the correct categories as the headings for the 2×2 tables, and if necessary change the headings, so that different aspects of the result can be considered. For example, in the field experiment, we tested the main result under the headings Abortions and Normal Pregnancies, but the heading

Normal Pregnancy may not have been strictly true. We must remember that the logical opposite of an abortion is a non-abortion, and that, although all abortions are abnormal pregnancies, not all abnormal pregnancies are abortions. If our recordings are good we shall have noted, not only the abortions, but the cows that have produced a living calf before the normal time (premature calving), those that have calved with such difficulty that it was necessary to give assistance (dystokia), and those that produced dead calves.

From our records, therefore, we might be able to rearrange our headings and record the following results:

	Abortions	Other Abnormal Pregnancies	Normal Pregnancies	Sub-totals
Treated	8	2	36	46
Controls	14	7	26	47
	22	9	62	93

We have now replaced our 4-cell by a 6-cell table. The ratios in the bottom sub-totals are now 22:9:62, a result that would be produced by the interference of two wedges, side by side, indicating two degrees of freedom.

The χ^2 test cannot be applied with safety where the theoretical number is as low as 5 except when we can use Yates' modification, but we cannot use this modification when there is more than one degree of freedom. We must therefore test to see if any theoretical value is as low as 5. Obviously such a low figure is likely to occur where the ratio in the bottom line is only 9 out of 93. It would also be more likely to occur in the smaller group, so that we can try $\frac{9}{93} \times 46$. This comes to 4·45 and the figure is therefore too small to test so we must find some other way of dealing with the table.

To increase the numbers in the cell with the low figure we can lump the two types of abnormality together, and this will give us the following 2×2 table.

	Abnormality	No Abnormality	
Treated	10	36	46
Controls	21	26	47
	31	62	93

$$\chi^2 = \frac{((260 \sim 756) - 46{\cdot}5)^2 \times 93}{31\ .\ 62\ .\ 46\ .\ 47}$$

$$= \frac{449{\cdot}5 \times 449{\cdot}5 \times 93}{31\ .\ 62\ .\ 46\ .\ 47}$$

$$= 4{\cdot}52$$

From χ^2 table p. ·05 = 3·84
p. ·02 = 5·41

It can be seen that this change of categories has altered the answer to the experiment. We have found that there was not sufficient evidence to justify the hypothesis that the treatment had decreased the number of abortions, but we have now found that there is evidence that the treatment had a beneficial effect and that further experimental work is justified to see if such a favourable result can be repeated.

Asking the correct question

This broadening of the category of abnormality is often a very important follow up when the main result is negative. It is particularly important in the case of diseases. For example, when using a vaccine, partial immunity may hide the clinical symptoms usually associated with the disease. Or, with control animals, the ill health produced by the disease may permit some other infection or pathological process to affect the animal and obscure the expected symptoms. It is always best to include an examination of Total Deaths or of Total Abnormalities in the analysis of the results, as well as looking round for other classifications.

To emphasise the care that must be taken in selecting headings the following examples may be of interest.

The first is of a worker on sheep diseases, who was very pleased with an attenuated virus he had produced to protect sheep from a certain disease, and as evidence he produced the following figures:

	Death from Virus Disease	Survivors and Deaths from Other Causes	
Treated	2	64	66
Controls	14	52	66
	16	116	132

Do you find that this result is significant?

The results were shown to an Important Person who was delighted and immediately gave orders that his own flocks should be innoculated. The results were disastrous. Some time after when the worker had moved to another job, someone went through his records for the individual sheep and produced the following figures:

	Death from X Disease	Death from Other Causes	Survivors	Sub-totals
Inoculated	2	27	37	66
Controls	14	7	45	66
	16	34	82	132

Looking at these figures we see that we now have a 6-cell table in which the different types of events are in the ratio 16:34:82. As there are two degrees of freedom we cannot apply the χ^2 test until we have seen that the lowest theoretical value is over 5. We have done this before; would you agree that it is done by multiplying the smallest sub-total in the bottom row by the smallest sub-total in the margin and then dividing by the total? In this table it comes to 8 and therefore we can go ahead with the test.

As there are two degrees of freedom we cannot use the quick formula but must find $\frac{(A-T)^2}{T}$ for each cell. The theoretical values for each cell are easily found for, as there are equal numbers in each group, then half the bottom line must be distributed into each group, giving 8:17:41. We cannot make Yates' modification because we have more than 1 degree of freedom, so that the value for the top left hand cell will be $\frac{(2-8)^2}{8}$ or 4·5. If you work out the other 5 in the same way, you will find the total for χ^2 is 21·54. Looking this value up in our table of χ^2 with $n = 2$ for the 2 degrees of freedom, we find the where $\chi^2 = 13{\cdot}8$, then $p = 0{\cdot}001$ so that having added this extra class the result is still highly significant, to such an extent that it falls outside the table.

What does this result tell us? All it tells us is that, if 2 wedges were arranged in such a way that they would, on the average, distribute shot in the proportions 16:34:82, it would be very unlikely for 66 shots, poured down on two consecutive occasions, to give the results shown in the table.

So far we have carried out two tests. First we compared deaths from *X* disease with other deaths and survivors, and the answer calculated from this should be $\chi^2 = 8{\cdot}6$, or highly significant. In the second test we had Deaths from *X* Disease, Other Deaths, and Survivors, and again it was highly significant.

If we are too frightened of statistical tests to think, or if we feel that these tests do our thinking for us, we might just accept that this second answer confirms the first. On the other hand if we accept these tests as a convenient way of displaying figures for our inspection, then we will notice that the 2 columns for Deaths from *X* Disease and Deaths from Other Causes, contradict each other.

Having noticed this we can, using the second table, ask a much more important question from the flock-master's point of view. That is, does the vaccine increase the number of survivors? The figures for this are:

	Deaths	*Survivals*	
Inoculated	29	37	66
Controls	21	45	66
	50	82	132

Here $\chi^2 = 1{\cdot}58$ which is nowhere near significance.

This, then, is an example of a weak live vaccine not being weak enough, making its recipients more susceptible to other diseases.

As another example of testing under different headings to obtain a true picture of the result, let us assume that a worker claims that he has found a fairly effective treatment that will lower the death rate from *Z* disease. He produces the following figures, pointing out that the death rate in control animals is three times as high as in the treated animals, and asks for financial support:

	Deaths from Z Disease	*Survived or Other Deaths*	
Full Treatment	3	171	174
Incomplete Treatment	7	33	40
Controls	10	194	204
	20	398	418

In this table there is only 1 wedge, and this divides the shot in the proportion of 20:398, but in this instance there is an extra batch of shot to be poured down the funnel, so that an extra comparison is provided, giving a total of 2 degrees of freedom. If we check the smallest theoretical value we will have $\frac{20 \times 40}{418}$, which is obviously less than 5, so that it is unsafe to treat this table as it stands, and we must turn it into a 4-cell table.

As the worker has only claimed results for the full treatment, we can start by comparing the full treatment with incomplete treatment and no treatment, giving the following figures:

	Deaths From Z	*Survived and Other Deaths*
Full Treatment	3	171
Not full treatment	17	227

The total and sub-totals will remain the same, and we see that $\chi^2 = 5{\cdot}02$. So that the result is just over significance, with a probability of about 3 in 100.

If statistics is our master we might accept this as proving the worker's results, and pay him the money. But if we regard statistical significance as just a useful yardstick, then we will use it to carry on our inspection. We might continue testing by comparing any treatment at all with no treatment at all. This would give:

	Deaths From Z	*Survivors and Other Deaths*
Any degree of treatment	10	204
Controls	10	194

Inspection alone shows that this is not significant, so that we can confirm that it must be the complete treatment that is necessary. Are we now satisfied? And do we pay up? Or do we test the remaining alternative of incomplete treatment against both full treatment and no treatment? This does not appear to be a very logical test, but let us carry it out for the sake of completeness. The figures are:

	Deaths From Z	*Survived and Other Deaths*
Incomplete Treatment	7	33
Full treatment or none	13	365

Do you agree that this gives $\chi^2 = 12{\cdot}76$? This is such a high value that it is outside Fisher and Yates' table, and the probability is less than 1 in 1000.

With a value like that it is very unlikely indeed that the higher death rate in the incomplete treatment group was due to chance alone. What was it due to? Was it because incomplete treatment made animals more susceptible to the disease? Or is it more likely that directly the investigator heard that an animal in the treated group was dead, he hurried round and closely questioned the attendant, and on several occasions was able to get an admission that the complete treatment had not been given? With either alternative we keep our money!

Summary

We have had a look at a field experiment and have seen that this is not something that can be planned sitting in a laboratory, and then left for someone else to carry out. If you are in charge then, to obtain a reliable result, you must make yourself responsible for the avoidance of all the errors; thus if a cattleman feeds the wrong cow, or if an animal dies and no one tells you, it is your fault. These faults are avoided by making regular visits, giving lucid instructions, and by taking a personal interest in the animals.

If results can be judged on the " Passed " or " Failed " basis, then the Chi-squared test will give a good approximation of the exact test results, where this test would be tedious because of the large numbers involved. Where the numbers in both groups are equal, the answers can be looked up in the tables prepared by Mainland, Hererra and Sutcliffe. More can be learned about the Chi-squared test by consulting one of the books on applied statistics, as given in the bibliography, particularly one with a medical bias, such as Mainland or Hill.

It is important to remember that these tests are merely tools, and that if you ask them a silly question they give a silly answer; this is not their fault, it is yours.

References

Topley, W. W. C., and Wilson, G. S. (1941). *The Principles of Bacteriology and Immunity*. Williams and Wilkins Co. Baltimore, p. 1050.

Chapter 3

THE NORMAL RANGE AND THE STANDARD DEVIATION

> There is something very sad in the disparity between our passion for figures and our ability to make use of them once they are in our hands.
>
> M. J. Moroney

So far we have dealt with contingency tables in which the number of things that " pass " or " fail " are recorded. Now we turn to a different type of data in which every individual has a class, or measurement, of its own.

The human mind dislikes anything to be indefinite, thus Walter Lippman refers to " . . . our human propensity to insist on having an opinion when all that we are entitled to is an open mind." The idea that every packet of sugar, clearly marked 1 Lb., really has its own individual weight, which is neither the same as that of any other packet nor exactly 1 Lb., is hard for the mind to accept, but, even using the kitchen scales, one can sometimes show that even the same packet will weigh under 1 Lb. in dry weather and over 1 Lb. in wet weather.

The dislike of indefinite statements, such as " about 1 Lb." or " approximately 1 Lb.", leads people in a search for the " true " average, and they feel much happier if they can state that the average packet of sugar weighs 1·063 Lb. Even this statement may lead someone else to use a more delicate method of weighing, so that he may produce a " better " average of 1·06274 Lb. This obsession with averages is seen everywhere, from statements that the average American High School Girl spends $1.375 on cosmetics every Wednesday in summer, to the statement that the average temperature of the rabbit is 100·85° F.

When we try to use these perfect averages we find that they are not much help to us. The average body temperature of the rabbit may be 100·85° but if we take the temperatures of those in our own laboratory we may find temperatures of 101·0, 101·6 and 101·8. What we want to know is are all these three abnormal? Are any of them abnormal? Have we been overheating the rabbit-house? All we can say is that all these temperatures are above the average, but what we want to know is: how far from the average can a temperature be without the animal being regarded as abnormal?

To get some idea of how to deal with the normal range let us change to the body temperature of the horse. As most text-books still limit their information to giving the average, we may begin by finding the average for the horse recorded as 100·0° F. in one text-book and as 100·5° in another. These figures give us little help in deciding abnormality and to find the normal range we would have to take the temperatures of a number of horses—even though people have been doing this for over 100 years. Let us suppose that we have collected the temperatures for 37 horses, and that these are as follows:

100·0	99·8	100·0	99·6	98·0	100·0
100·6	99·8	100·0	100·0	99·0	101·4
100·0	99·6	99·8	99·8	100·0	99·4
100·0	101·0	99·8	99·6	99·4	100·2
99·6	100·4	101·4	100·6	99·8	100·4
99·8	100·4	100·0	98·4	98·8	100·6
					100·0

Average 99·919° F.

The first thing we might notice is that this average is not the same as either of those recorded in the text-books. Secondly, we might see that none of the horses actually had the average body temperature. Thirdly, we can see that 1 horse was as low as 98·0° and another as high as 101·4°; were these 2 animals healthy? The only answer is that they appeared to be healthy, and that they were doing a good day's work.

What can be done with collections of figures like these in order to use them with ease? According to Mather the object of statistical analysis is to reduce bulky and complex data, that the mind cannot fully comprehend, to a few easily understood quantities that contain most of the relevant information. However much you may dislike the idea of using statistical methods, if you try to simplify these figures in any way you are using such a method. It may be a simple method worthy of a child of ten, or it may be a mature and efficient method, but which ever it is, it is a statistical method.

To consider a simple method of concentrating figures, imagine a man, far back in history, living on a tropical island, and suppose that he wanted to convey some idea of his coconut crop to a possible buyer, by taking him a sample.

If this simple soul could only carry a sample of about half-a-dozen nuts he might take the first 6 that came to hand. If he existed even before the proverb about honesty being the best policy, he might select 6 big ones, but if he was a man of genius he could have proceeded as follows. First he could put the coconuts in a long row, with the smallest at one end and the largest at the other. This is called putting them into an array, and from this he would be able to note the smallest one (the minimum) and the largest one (the maximum). Next he could find the middle one, or median. This would be a good start but he could go one important step farther. As he had found the middle one he could split the whole array into halves and then, treating each separately, he could find a median for each half, and as these divided the array into 4 equal parts these markers could be called quartiles.

By selecting these 5 specified nuts he could go to the buyer and say " My crop consists of as many nuts as there are pebbles in this bag. The smallest nut is this size and the largest is this size. Half the crop is between this size and this size with the middle one this size." By doing this he could convey far more information about his collection of measurements than is given in some technical articles even today.

Note that if it was a long journey and he wanted to carry only 2 nuts, he could take the 2 quartiles. From these he could barter on the fact that half the nuts lay between these two sizes, and that those outside the quartile at one end would roughly compensate for those at the other end. Alternatively, if he wanted to take a larger sample, and had, for example, 400 nuts in his crop, he could divide them into percentiles. With this method selecting every fortieth nut would give him 10 percentiles, or deciles.

Inefficiency of the maximum and the minimum

It has been suggested that if he were to restrict himself to 2 nuts to indicate the size of the nuts, he would choose the quartiles. Why should he not select the maximum and the minimum? The reason is that these extremes are often freaks; the greater the chance combination of favourable factors the bigger the maximum; the greater the chance combination of unfavourable factors the smaller the minimum.

The fact that chance plays a big part in producing these extremes also encourages personal interference to modify the effect of chance, and this interference, in science, is unforgivable. The owner of the nut crop may look at his array and say to himself " This smallest nut is so much smaller than all the others that the more I look at it, the more convinced I am, that it must be diseased. It would not be fair to include a diseased nut with my sample, so out it goes." Looking again, he might think that the largest nut was unusually small that year. This would be unfair because it suggested that the whole crop was smaller than usual. If there was a tree, just outside his estate, bearing a single large nut, would it not give a fairer picture of his crop if this one was added as the maximum? Note that this artful manipulation would make little difference to the quartiles, because all it would mean was that they were each moved one nut up.

In our islander the stimulus for the interference was the profit motive; in scientists it can be vanity, or honest ignorance. With vanity both extremes may be removed for fear that colleagues might think he made larger technical errors than they did. With honest ignorance one or both extremes might be removed because they seemed so extreme that it was felt that they indicated ill-health, or a mistake in calculation. Reading technical articles with an experienced and disillusioned eye, one can detect unnaturally close distributions being carried on by one worker after another, until they are broken by someone broadening the distribution by using the standard deviation as the estimate of dispersion.

Depicting a collection of data

It is useful to be able to depict the pattern the data form by showing the measurements in a diagram. If we return to the sample of temperatures we can see how these can be depicted as an array by looking at Fig. 47. The middle temperature is one at

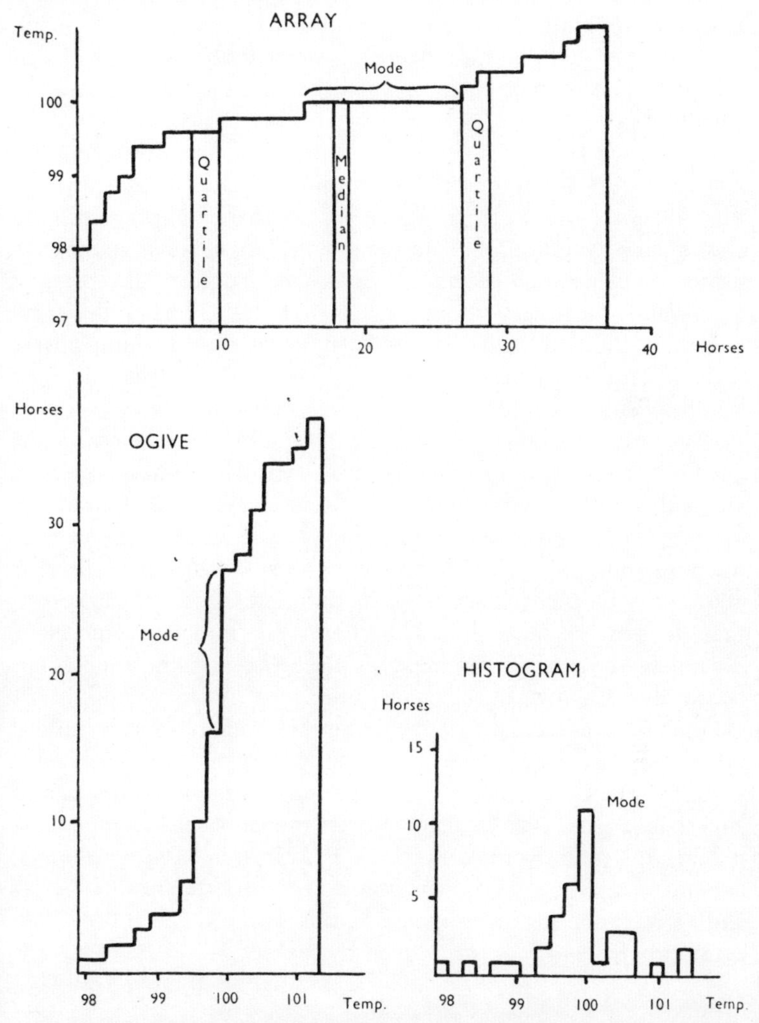

FIG. 47. Methods of depicting a sample of 37 body temperatures.

100°. Removing this we have 18 measurements, an even number, on each side. This means that the first quartile will fall between the eighth and ninth item, but as both these are 99·6° the first quartile will be 99·6°. The third quartile is between 100·2 and 100·4, so that we record 100·3°. Note that, because there are more temperatures at 100° than at any other, the 100° measurement is known as the mode. We could use this array to tell other people that half the temperatures were between 99·6° and 100·3° and this would be far more help to them than the fact that the average was 99·9°.

If we turn the array round so that the 37 measurements run from top to bottom, instead of from side to side, as shown in Fig. 47, they will form an ogive. With these temperatures this shows no advantage over an array, but an ogive is sometimes useful when we wish to show time or dose from left to right, and death rate, or infection rate, from bottom to top. Thus if this ogive did show the death rate, it would show that all 37 horses were dead when the temperature rose to 101·4°. Note that this gives a slightly *S* shaped curve.

The best method of discovering the pattern produced by the data is the histogram, and this is also a much quicker method when there are a large number of items. In the histogram we are prepared, if necessary, to sacrifice the strict accuracy of each measurement. We divide the base line so that, where the data allow, it will be divided into 20 different classes, arranged so that, in general, several measurements fall into the same class. To represent each measurement we place a brick in the appropriate class; this is shown in the same figure.

One of the advantages of the histogram is that you can build it up as you increase the size of the sample, and also in each brick you can put the name, or number, of the horse it represents. You could go even further, and introduce such variations as putting the mares' numbers in ink and the geldings' numbers in pencil. Further you could draw a red brick for young horses and a green one for older ones. In this way you could keep your eyes open for any bias introduced by sex or age.

The biological distribution

In spite of the fact that there are only 37 items in the sample of temperatures, by displaying them as a histogram there is a rough indication that the mode and the median are near the middle of the temperature classes, and that the items fall away in numbers each side of the mode. It does not require much imagination to accept that with a larger sample the items would give a hump in the middle and fall off in numbers on each side to form a shape like an admiral's hat. This type of distribution is therefore called a cocked-hat curve and, because such a shape is so often formed by biological material, we need to know more about it.

It will be seen that by sacrificing exact measurements and putting the items into equal sized classes along the base of the histogram, the items give a shape very similar to the frequencies given by the Pearson's wedges, used to demonstrate the binomial distribution. If we over-simplify the whole problem of the

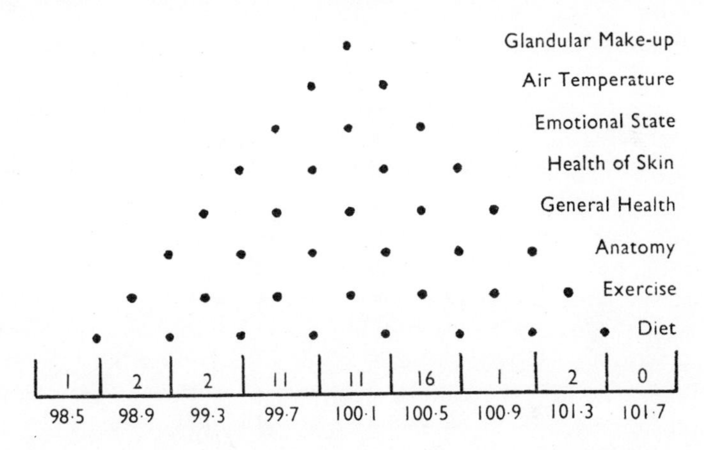

FIG. 49. Heredity and environment as causes of variation in body temperature.

physiological variation of temperature we can use Fig. 49 as an illustration of how such shapes are formed.

Firstly we may believe that part of the system controlling body temperature was the glandular make-up of the animal. In the very complex system of endocrine glands the relative size and functional activity of each will be partly inherited, and dependent on the chance absence, or presence, of different genes. In our working model the effect of this whole complicated interaction is shown by a single wedge, and if the bias of the glands is to run the body temperature at above average, the shot will be thrown to the right. The next wedge represents the temperature of the surrounding air, with variations of the animal being heated by sunlight or cooled by the wind; the next represents the emotional state, including the ability of anxiety to increase the temperature. Other wedges represent health, anatomy including the nearness of the blood vessels to the surface of the skin, exercise, and diet.

Each of these very complex, multiple, factors has been allocated only a single wedge in the figure, with the result that we have distributed the temperatures into about 8 cells. Note that for simplicity one temperature has been omitted. The temperatures run in a series 98·4, 98·6 then 98·8, 99·0, and so on. Where we have more than one measurement in a cell the cell is labelled with the middle value, thus 98·4, 98·6 are both recorded as 98·5. Because these cells, or classes, increase in little steps instead of a continuous line, this type of distribution is called a " discontinuous " one.

The normal curve of error

If several thousand horses had been sampled, and if a much more sensitive method of measuring temperature had been used so that temperatures could be graded as 98·001, 98·002 and so on, then because each governing factor is really equivalent to many lines of wedges instead of just one, we would expect a distribution in which the steps between one class and another would be so small as to be nearly invisible.

Where we have a binomial distribution in which a very large number of wedges is based on equal odds, for or against, it is found that a line passing through the tiny steps of the histogram gives the shape of a church bell. This shape is the basis for a great deal of the knowledge of statistics, it is illustrated in nearly every book on the subject, and it is known as the normal curve of error, or the Gaussian curve, the latter because it was formulated by Gauss when he was investigating the multiple errors present in the calculation of the position of various planets.

A normal curve has various measurements, such as the average, the median, the quartiles, and the standard deviation, by which its size and shape can be calculated. These measurements are known as parameters, and the parameter we are interested in, in order to indicate the range of measurements, is the standard deviation.

Just in case you feel that to talk of the normal curve of error is leading you into modern advanced mathematics, may I point out that it was first discovered by De Moivre (Wallis and Roberts) who was a friend of Sir Isaac Newton and a fellow member of the Royal Society. It was some years after De Moivre's death that James Watt discovered how to make a practical steam engine.

The standard deviation

The quartile is a useful measurement, but it would be tedious to use it on a sample of, say, 300 items if it was first necessary to display them as an array. Statisticians therefore tried to find some method of calculating the quartile, and with the formulation of the normal curve it was found possible to calculate the quartile by first calculating the standard deviation, and then multiplying this by 0·67. It was soon recognised, however, that the standard deviation itself was an even better index for measuring the range.

To see what the standard deviation (s.d.) is, and how it is calculated, let us take a sample of 6 items, measuring 1, 2, 3, 3, 4, 5, and deal with these by the method shown below:

Items	*Average*	*Diff.*	*Diff.*^{2}	
1	−3	−2	4	Mean square (or variance) $\frac{10}{6-1} = 2$
2	−3	−1	1	
3	−3	0	0	
3	−3	0	0	Standard deviation $= \sqrt{2} = 1{\cdot}414$
4	−3	1	1	
5	−3	2	4	
18		0	10	Sum of squares

The items, for convenience in an array, are shown in the left hand column. They total 18 for the 6, giving an average of 3. This average is subtracted from each term, hence if the arithmetic of the difference column is correct it will add up to 0. In the next column these deviations from the mean are squared and thus lose their minus signs; thus $-2 \times -2 = +4$. These squared differences are added together to give the sum of the squares of the deviations from the mean.

To find the mean square, or variance, the sum of squares is divided, not by the number of items, but by the degrees of freedom. We have already said that, if great accuracy of measurement is used, no two items are of exactly the same measurement, so that they can be put into an array. To obtain an array of 6 items only 5 wedges are required, so that we have 5 degrees of freedom, and the variance is $\frac{10}{5}$, or 2. To find the standard deviation from this we take the square root, thus the s.d. $= \sqrt{\text{variance}} = \sqrt{2} =$ 1·414.

ROUTINE SHORT CUTS IN FINDING THE STANDARD DEVIATION

Before we see how the s.d. is used, let us try to find it on the sample 37 temperatures. These are awkward figures to deal with and they permit the opportunity of demonstrating one or two short cuts. We could, if we wished, proceed exactly in the same way as we did above, but we found the average temperature to be 99·919° and it would be tedious to subtract this from each item, so we use a false mean. Secondly, the difference between each class is 0·2°, and if we square this it is 0·04° and it is rather hard to remember where the decimal point is, we therefore use a false deviation. If you find these short cuts difficult, remember that you need not bother with them until you think that they might be useful to you. The different steps are given in Table 53.

In the first column we write down the complete range of temperatures, moving in equal steps of 0·2°. In the second column we record the number of horses showing that temperature. Together these two columns form a frequency table. Note that the symbol *df* stands for deviation times frequency and do not get misled into thinking it has any connection with degrees of freedom.

Instead of beginning by calculating the true average, we have chosen a false, or working average that would be convenient. In this instance 100 is easy to subtract particularly as this number is also the mode.

With 100·0° as the working mean, or the false mean, then 99·8° is one division below it, and the difference would be −0·2°. This is an awkward figure to deal with so we will call this difference one division below the working mean. With a working mean of 100·0 and the difference between each temperature called a division, we can fill up column 3 which gives the number of divisions each temperature is away from the false mean.

TABLE 53

Finding the standard deviation for the temperatures of 37 farm horses working in the same district during the same month.

Temperature	*Frequency* (f)	*Deviation* (d)	df	d^2f	
98·0	1	−10	−10	100	
2	0	− 9	0	0	
4	1	− 8	− 8	64	
6	0	− 7	0	0	
8	1	− 6	− 6	36	
99·0	1	− 5	− 5	25	
2	0	− 4	0	0	
4	2	− 3	− 6	18	
6	4	− 2	− 8	16	
8	7	− 1	− 7	7	
100·0	10	0	0	0	
2	1	1	1	1	
4	3	2	6	12	
6	3	3	9	27	
8	0	4	0	0	
101·0	1	5	5	25	
2	0	6	0	0	
4	2	7	14	98	
			−50	429	False sum of squares in divisions
			35		
	Total error in divisions		−15		

One division = 0·2° F.

Correction of Mean (or Average).

Total error = −15 divisions

Average error = $\dfrac{-15}{37}$ = −0·4054 div.

Average error in degrees = −0·4054 × 0·2 = −0·08108°

True average = False average + average error

= 100° + (−0·08108) = 99·92° F.

Correction of Sum of Squares

Correction term = $\dfrac{(\text{Total error})^2}{\text{Items}} = \dfrac{(-15)^2}{37}$

= 6·081 divisions

True sum of squares = False sum of squares—correction term

= 429 − 6·081

= 422·9 divisions

Standard Deviation

Mean square (or variance) $= \dfrac{\text{Sum of squares}}{\text{Items} - 1} = \dfrac{422 \cdot 9}{37-1}$ = 11·75 div.

Standard deviation = $\sqrt{\text{variance}} = \sqrt{11 \cdot 75}$ = 3·428 div.

Standard deviation in degrees = 3·428 × 0·2 = 0·6856° F.

RESULT: Mean 99·92° F. with S.D. of 0·6856°
or approx. 99·9° with S.D. of 0·7°.

There are several horses at some of these temperatures so that, instead of repeating the difference for each horse under each other, we multiply the difference, or deviation, by the number of horses in which it occurs, this gives us the fourth column.

We have now to square the deviations, and this can be done either by multiplying the fourth column by the third column, or by squaring the third column and multiplying it by the second column. These figures when added give us the sum of squares for the divisions away from the false mean.

We have obtained the false sum of squares very easily but now we must pay for the short cuts we have used. The first question is what is the true mean? If we had used the true mean the column "df" would have come to zero when totalled; instead it leaves a residue of -15 divisions which must be spread over 37 items. As each division is of $0{\cdot}2°$ this gives us an average error in degrees of $\frac{-15 \times 0{\cdot}2,}{37}$ or $-0{\cdot}08108$. The correct average is the false average plus the average error, or $100°-0{\cdot}08108°$ which is near enough $99{\cdot}92°$ F.

Through using a false mean the sum of squares is too big. We remove the excess by subtracting a correction factor that consists of the total error squared, divided by the number of items, and this is $\frac{(-15)^2}{37}$ or $\frac{225}{37}$. Note that squaring has removed the minus sign, which is correct, for whether the working mean is too high, or too low, it makes the sum of squares more than it should be. Subtracting the correction factor from the false sum of squares we have $429-6{\cdot}081 =$ about $422{\cdot}9$ divisions for the true sum of squares.

To find the variance we divide the sum of squares by the degrees of freedom which we know is $37-1$. This gives $\frac{422{\cdot}9}{36}$ or $11{\cdot}75$ divisions. The standard deviation is the square root of this, which is $3{\cdot}428$ divisions. As each division is only $0{\cdot}2°$ we must multiply by $0{\cdot}2$, giving a standard deviation of $0{\cdot}6856°$ F. So that with both average and standard deviation back in degrees we have a result of $99{\cdot}92°$ F. with an s.d. of $0{\cdot}6856$.

Geometrical method of finding the standard deviation

If you find it difficult to understand the use of the various short cuts, or if you dislike any type of calculation, there is still a way out, for you can find the standard deviation by using a geometrical method. This method was introduced by Woolf (1949) and all

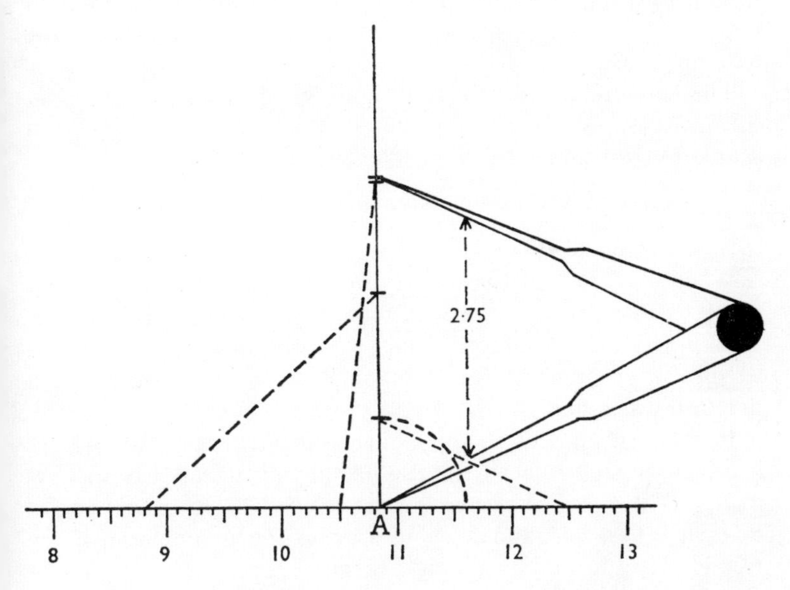

Fig. 55. Geometrical method of finding the standard deviation.

that is required is an efficient pair of dividers and some good quality squared paper.

To demonstrate this method on a simple example, assume that we wish to find the s.d. for four measurements, viz., 11·6, 12·5, 8·8, and 10·5. These figures give an average of 10·85.

First draw a base line on the squared paper, and using a suitable scale, mark in, as accurately as possible, a perpendicular which touches the base at the average of 10·85. Call this point *A*. Place one point of the dividers accurately on *A* and extend the other point to touch the first measurement of 11·6 on the scale. Keeping one point on *A*, rotate the dividers to transfer this measurement to the perpendicular (see Fig. 55).

Keep one end of the dividers on this point on the perpendicular, and draw out the other leg to touch the second measurement of 12·5 on the scale. Lift the dividers and placing one point on *A* prick the other point onto the perpendicular. Keeping one leg there, again draw out the lower leg to touch the next measurement, 10·5, and again transfer this distance to the perpendicular. Repeat with 8·8. Measure the distance from this highest-point on the perpendicular to *A* on the base line, and measure it against the base line scale.

The measurement for the four items was 2·75, so we now apply the formula:

$$\frac{\text{Total distance on perpendicular}}{\sqrt{\text{Number of items minus one}}} = \frac{2{\cdot}75}{\sqrt{4-1}} = 1{\cdot}59.$$

Result, average 10·85 with s.d. 1·59.

When used to find the s.d. of the 37 temperatures the result was 0·666 instead of the s.d. of 0·6856 given by the calculation. With practice the accuracy would be greater, but even so it would be difficult to imagine that this small error would make much practical difference.

If, when using this method, the perpendicular gets too long, it can be parked somewhere else on the paper, and a fresh start made. These parked perpendiculars can be plotted at right angles to each other, and added by measuring the hypotenuse of the triangle they form with another sheet of paper. If you wish to use this method later on for an analysis of variance, or for regressions, then you should consult the original article.

The t distribution

We have seen how to find the standard deviation, and we have said that with the normal curve of error we know that 0·67 of an s.d. will be equivalent to a quartile, that is, limits of the mean plus 0·67 s.d. and the mean minus 0·67 s.d., will enclose half the population. The limits set by other quantities of the s.d. are given in nearly every book on statistics, as well as in statistical tables (e.g. Fisher and Yates) but it is easy to remember that about two-thirds of the population is contained by 1 s.d., 19 out of 20 by 2 s.d. and 399 out of 400 by 3 s.d. Thus the 2 s.d. limit lent itself as a suitable arbitrary limit for significance.

The normal curve is measured by exact parameters because it contains the whole population. In biology, to obtain the distribution for the body temperatures of horses we would have to measure the whole population, that is, every healthy horse in that environment, and it would take several hundred horses to produce the curve. With much biological material it would be too expensive, or even impossible, to obtain such a large number of measurements and we have to be content with a sample of the population. Thus, instead of the true mean, we have the mean of a sample, and instead of the true s.d. we have the s.d. of the sample. These measurements when based on a sample rather than on the population are spoken of as "statistics" and not parameters.

"Small samples" says Moroney, "are slippery customers whose word is not to be taken as gospel." To demonstrate this, assume that out of 100 items 1 is much the biggest, then if we divide these into 10 samples of 10, 9 of these samples can have a smaller range than the original group of 100. To be prudent in judging any item to be abnormal we take two precautions. First, as we saw, when finding the s.d. we divide the sum of squares by $n-1$, instead of n, or the full number of items. Secondly, we look up the value of the s.d., not in the table for the normal curve, but in "Student's" table of "t," where an allowance has been made to compensate for the size of the sample.

Returning to the sample of temperatures, in which we found that the mean was about 99·92° and the s.d. about 0·69, we can look up the table of t in Fisher and Yates' tables. We see various probabilities along the top, and the number of degrees of freedom run down the side. (See appendix). For 37 temperatures the degrees of freedom will be 36, but as we cannot find 36 we must make a choice between either 30 or 40; to be cautious we choose 30. We may then look down the column under $p = 0{\cdot}05$, the equivalent of 1 in 20, given by 2 s.d. in the normal curve, but in the t table we find opposite 30 the figure 2·04. From this estimate we judge that if we went on increasing the size of our sample, 19 out of 20, or 95 per cent of temperatures, would fall within the limits of $99{\cdot}92+(0{\cdot}69\times 2{\cdot}04)$ and $99{\cdot}92-(0{\cdot}69\times 2{\cdot}04)$; that is, between 98·51 and 101·33. For $p = 0{\cdot}01$, where 99 per cent fall within the limits, we find 2·75 s.d. Thus if we increased the sample we would expect to find 99 per cent. of the fresh items falling between 98·02 and 101·82° F.

The normal range

We have seen that, because our sample of 37 temperatures produced a histogram that suggested that the sample came from a cocked-hat distribution, we were able to use the s.d. to estimate the proportions in which different temperatures would occur in future samples. This knowledge would not be very important under these circumstances, but if the circumstances were different it would help, say, a clothing manufacturer, to meet his future requirements. Thus he would know that he could satisfy 2 out of 3 customers if he made only those sizes which would fit those measurements that fell 1 s.d. each side of the mean, and that there would be only one-twentieth of the total demand for sizes that would fit measurements more than two s.d. from the mean.

We can, however, use our sample for two other purposes. Firstly, we can compare our sample with other samples, collected by other workers, elsewhere, to discover whether there are differences due to breed, or to environment; this will be dealt with in the next chapter. And secondly, we can use our sample as a guide to the normal range of temperatures for healthy horses, and here we are on less orthodox ground.

In using estimates based on the normal curve to predict biological material, we have a safeguard based on the fact that biological material must be in harmony with its environment. In a pack of playing cards the possible ways of arranging the 52 cards is factorial 52, and this is more than a million million. Thus the chance of a pack being in any order is less than one in a billion, nevertheless the pack is always in some order or another. If we used the normal curve to predict horses' temperatures our sample might suggest that once in a billion times we would have a horse with its blood at freezing point, or one with its blood at boiling point. Neither of these horses would be happy horses, in fact they just could not exist. We see, therefore, that the two ends of cocked-hat curves formed by biological material are cut off more abruptly than with inanimate objects that depend on the laws of chance alone.

When, however, we try to set the limit between health and disease, we find that we are left very much to our own resources, for this is a most difficult task for either words or figures. When you add to this the fact that we cannot be sure that our sample is from a perfectly normal curve, it can be seen that we must not expect too much from our criterion, for arithmetic cannot decide accurately what intelligence and experience have failed to do. The fact remains that it is often necessary to decide on some arbitrary criterion as a limit to health, or suspicious of disease, and although not mathematically correct, the s.d. appears to be the best method of estimating it. To take the ordinary criterion of significance of $p = 0{\cdot}05$, and use this as a 95 per cent. confidence level, condemning the 5 per cent. that fall outside it to be of doubtful health, is too drastic.

The arbitrary limit must therefore be decided by each worker from his own experience of his material. In my own experience I find it helpful to begin by the assumption that any measurement more than 2·5 s.d. away from the mean is suspicious of disease, and that measurements 3s.d. away from the mean are definitely abnormal, that is, the animal is diseased or there is a gross error in technique.

If we apply these criteria to the samples of temperatures, where the mean was 99·92° and the s.d. was about 0·686°, we have $99{\cdot}92 \mp 2{\cdot}5 \times 0{\cdot}686$, which gives limits of 98·205 and 101·635, so that temperatures of 98·0° and 101·8° would be considered suspicious. We see that these limits cut out one horse which has a temperature of 98·0°. Any additional horse at this temperature would be viewed with suspicion, and if possible, the temperature would be taken again to check for technical error, for although this error must be kept included when making the range, we need not include it when trying to determine if an animal is ill.

Some different types of distribution

We have seen that the main thing we wish to know about a sample is its shape, rather than the extreme values that our random sample has fished up. We have learned that, with much biological material the shape tends to be that of a humped curve, or cocked hat, which can be described by using a statistic called the standard deviation; a good method of depicting the shape being to build a histogram with the items.

As one can still read articles in technical journals where samples are described by detailing the mean, the minimum and the maximum, let us see how inadequate these statistics are, by seeing how much more is conveyed by putting the items into a histogram. Thus, if we are told that a sample of 20 items has a mean of 8, a minimum of 3 and a maximum of 12, what shape does it have? Some of the possible shapes are depicted in Fig. 60.

Histogram A in Fig. 60 represents the type of distribution usually found in biological material. The mean is 8 but the mode is at 9,

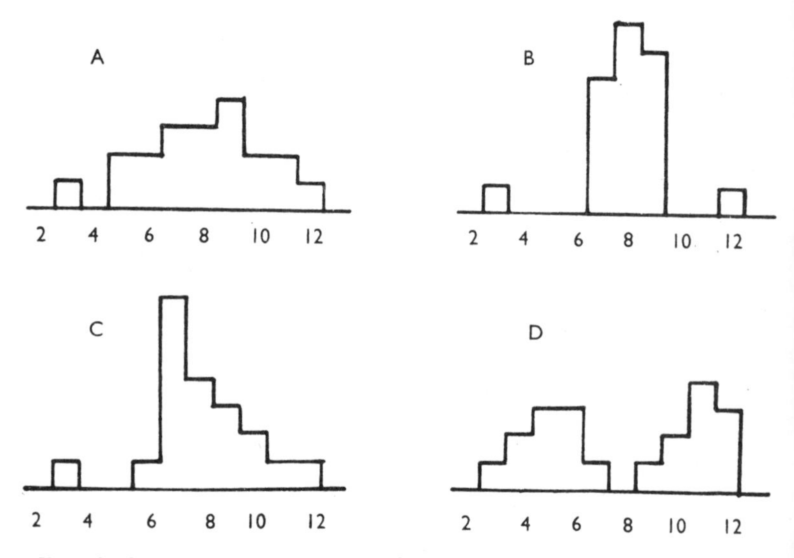

FIG. 60. Four possible distributions of 20 items each with average 8, minimum 3, and maximum 12.

otherwise it is a cocked-hat curve, with a lot of items near the mean and their numbers becoming fewer as the items become more extreme. Additional items should make the mode and the average fall into the same class; if not, the distribution will not be a perfect cocked-hat curve but would be slightly skewed to the left; skewness is depicted more clearly in histogram C.

Histogram B shows a much more condensed distribution, standing mainly over units 7, 8, and 9. The most notable feature is the two isolated extreme items; and every beginner in research work will itch to cross them off. If he lacks the courage to cross them off he might go back to the original recordings. If these have been made on a grubby piece of paper, with a blunt pencil or a ball point pen, then by staring at it long enough he may convince himself that, for example, what he thought was 98·0° was really 98·6°. Timidity is quite an important factor in wishing to cut off the unwanted, for the junior worker may believe that, if he presents a histogram similar to B, he will be told that he doesn't even known how to take a temperature. Similarly, seniors who reprimand subordinates for producing histograms like B will never be presented with them, nor will they ever be presented with much of the truth about anything.

If all the items in histogram B come from the same population, or universe—that is, have come down the same set of wedges—then, as more items are added to the sample, the empty spaces in B may begin to fill up to give the usual biological cocked-hat curve. Other alternatives are, first that the 2 extreme items are freaks, perhaps pathological in character, and secondly, that the sample contains items that have come down 3 different sets of wedges. For example, the main sample may represent mature healthy animals, the item on the left may build up into a population of unweaned animals of the same breed, and the one on the right build up into animals affected by old age.

If you are timid it is best to realise that the indignity of being accused of poor technique is less than having to eat your own words, which may easily occur when somebody else proves the existence of extreme values and comes to you for confirmation. It is well to remember that once you have written a paper proving extreme figures do not exist it is rather difficult to write another paper explaining their existence. It is surprising how often the item that you would like to push aside because it does not conform to the others turns out to be the most important measurement of the lot, and ultimately forms the basis of a fresh hypothesis; the important rule is a simple one: keep all the extreme results in your records, one day you may be glad of them.

Histogram C shows a distribution in which the mode is well on the left of the average, it can therefore be described as skewed to the right. This type of curve is seen when the chance of an event occurring, or not occurring, is far from equal; it resembles a wages curve with many at a low wage and only a few at a high one. It would be unlikely for a curve, so definitely skewed with 20 items, to turn into an ordinary biological distribution if further items were added.

With histogram D there are two modes; and no items fall on the average, suggesting pretty strongly that this is not the best statistic to use in describing the result. It is not a result that one would expect if all the items came down 1 set of wedges, and strongly suggests that the sample comes from 2 populations, each with its own mode and mean. It represents a challenge to the research worker to find out the attribute that is responsible for the difference.

In discussing these histograms I have tried to show that an important part of statistical method is to display the items in a sample so that their frequency gives a shape to which some hypothesis may be attached. To merely record the mean, maximum and minimum, entirely ignores the difference in the histograms, and this fact, together with the desire of some workers to conceal their extreme values through fear of error, make these values unsuitable; thus it is not surprising that Yule (Mainland 1938) described the maximum and minimum as the worst possible measure for any serious purpose.

Although extreme values should not be used to describe a sample without their being supported by percentiles, a frequency table, or a histogram, or some other statistics of shape, they are of interest from the " museum " point of view, and the highest, or lowest, temperatures ever recorded for a healthy horse is important physiological information, and some workers develop a collector's zeal for these extremes.

Building a sample

If our sample of temperatures takes the form of a survey then we should use the correct survey methods and make sure that our sample was representative of the whole population to be studied. Thus it would contain young and old horses; male, gelded, pregnant and non-pregnant mares; and horses from here and from there; and, secondly, it would have to be a random sample in that every horse in this population should have an equal chance of being chosen. These points are dealt with in a later chapter.

On the other hand our sample might be just any healthy horses that came into our hands. This would not be an ideal statistical method of obtaining a sample, but if we are always going to insist on the ideal conditions then the greater part of our research life will be spent in sitting down doing nothing. If we are building up a sample as horses become available, then all we can do to control conditions is to build up a detailed histogram as we go along, and to watch for bias; thus we may notice that red squares with a diagonal line are clustering on the right side of the distribution and this may tell us that young horses in May and June have a bias to higher temperatures. We must then decide if we are going to have this as a sample of healthy horses of all descriptions, or whether we are going to accept these young horses as a population of their own, and start a fresh histogram for them alone.

It might be that, instead of collecting temperatures alone, a worker may be noting the respiration rates, the pulse rates, the height and the weight as well. In these circumstances each animal can be recorded on a different card, coloured and cut according to the available information; these cards can then be built up into histograms for each attribute in turn.

Sample size

How many animals do we need in a sample? That depends on the distribution the items give. If it is a nice compact distribution, with the mode in the middle, then we can be guided by the table of t (See appendix or Fisher and Yates, Table III). In the table you will see that the 95 per cent confidence level for 6 items, giving 5 d.f., is 2·6 instead of 2 s.d. as in the normal curve (more accurately 1·96 s.d.). From 5 to 10 d.f. the reliability increases definitely with each additional item, and with 20 d.f. even the 99·9 per cent confidence level ($p = 0{\cdot}001$) has become fairly steady. When we come to 31 items the 99 per cent confidence level is only about 6 per cent greater than that for the normal curve ($p = 0{\cdot}01$, 2·750 cf. with 2·576) so that any further increase in the number of items has little value in fixing the range.

Where the distribution is not compact, and several gaps are left on the base line, there may be several unknown causes of bias.

In this event you must go on collecting samples until the shape becomes obvious or the cause of the bias is discovered. To get rid of the gaps where there is a large amount of variation in the items may require some 80 to 100 items. The fault with many workers is that they stop before they have seen what shape their sample has, and one can often see articles in which a worker tries out some new method of estimation on, say, rabbits, rats and cavies, using 10 of each, when it would have been far better if he had sampled one species adequately.

Multiple samples

If, during a play, a conquering army marches across the stage, you may notice a pikeman with bow legs, later you may notice an archer with bow legs, and later still a swordsman with bow legs. By this time you realise that it is the same man going round and round, changing his hat and his weapon in a wild scramble behind the scenes. This subterfuge is used to build up what is called a " stage army " and it gives the appearance of a large army of varied weapons, but alas! if called on to fight it could produce only a few men each with several different hats and weapons.

This type of thing can occur in sampling. A worker is allowed the experimental use of 10 rabbits. He samples them once and finds that the results give a very ragged histogram, so he samples them again and again until he has a nice looking histogram of about 100 items. He has produced a " stage sample ": as his items are not from different rabbits he cannot use the distribution to judge any additional rabbit. It resembles an election where, to save work, votes have only been given to every tenth person, but to make up the numbers these people have been given 10 votes each.

This question of multiple samples sometimes arises when one is sampling a small group each month to try to show a seasonal variation, and later, wish to show a normal range. Here there are two alternatives. One can show a normal range for each of the months, or one can make a random selection of one sample from each animal. Whichever way you do it, 10 animals provide 9 degrees of freedom, neither more nor less.

Summary

With many biological attributes the individual values depend on the interplay of numerous factors, some dependent on chances that govern genetical make up, and others dependent on the chance circumstances of environment. These numerous factors support, or cancel each other, to such an extent that they produce a distribution very similar to the normal curve of error about which much is known.

This distribution, which resembles a humped curve or a cocked hat, is best described by recording its average to mark the centre, and its standard deviation to measure the scatter, or range. And the proportion of the distribution excluded by different numbers of s.d.s can be looked up in the table of t.

To record only the average, maximum and minimum, is a poor method of describing a distribution, although the maximum and minimum may be of interest to collectors. Where a distribution is not a cocked-hat type the mode as well as the mean should be given, and the type of distribution should either be described, or detailed in a frequency table. A histogram that shows more than one mode, or is definitely skewed to one tail or the other, supplies a clue that some samples may belong to a different population; these samples may represent, for example, those taken from a different age group or those from animals suffering from disease.

Reference

Mainland, D. (1938) "The Treatment of Clinical and Laboratory Data" Oliver & Boyd, Edinburgh, page 136.

Woolf, B. (1949). " Rapid calculation of Standard Deviation." *Nature*, **164**, 360.

CHAPTER 4

SAMPLES, ERRORS AND VARIANCE

Without generalization there is no meaning, and without concreteness there is no significance.

We have seen that a collection of temperatures from the same population of horses gave a histogram that had the mode roughly in the middle, with less and less concentration of items towards each extremity. The large number of times this curve occurs in nature has led to it receiving many names. Of these names the normal curve of error, the Gaussian curve and the bell-shaped curve, usually refer to the mathematically perfect theoretical curve, while the humped back curve, the cocked-hat curve, and the biological distribution, usually refer to the more irregular shape given by the smaller number of items that can be sampled in actual practice. As the cocked hat is now rarely seen, and as most dictionaries give little help as regards its shape, it should, perhaps, be pointed out that it was the style worn by such well-known figures in history as Napoleon, Nelson and George Washington. It consists of a hat with a wide brim. The brim is pushed up to a vertical position on both sides, giving the hat a somewhat triangular outline that resembles a binomial distribution. In contrast, each side of the bell-shaped curve starts as a convex curve and then changes to a concave one, the standard deviation being positioned where these two curves join.

In the last chapter we saw that, because the sample of temperatures gave a roughly cocked-hat distribution, we could estimate a range of normality. We could then use this range to judge whether any additional temperature was likely to be normal or abnormal. It will now be shown that where the items give a rough cocked-hat curve we can judge how far the means would vary if we took other sets of samples from the same population and, further, once we have discovered the error in our technique, what proportion of the total variation is due to physiological variation alone.

The standard error

In the previous chapter it was stated that the healthy temperature of the horse was recorded in one text-book as 100·0° F. and in another, 100·5° F. Even a small experience of biological measurements will tell us that it is unlikely for there to be any real difference between our average of 99·92 and 100° F., and we would take one average as good confirmation of the other. But is there a real difference between 99·92 and 100·5? To judge this objectively we must calculate what Fisher called the standard error. Although Fisher's term of standard error will be used, many workers do not restrict the term to this statistic, and it is also referred to as the standard deviation of the mean, or even as the standard error of the mean.

In trying to judge if one average comes from the same population as another, there are two factors that help the comparison. First, the averages taken from different sets of samples from the same population form a cocked-hat curve, even if the whole population does not, itself, give a cocked-hat curve. And, second, the larger the number of items in a set of samples, the nearer the average of this set will be to the average of the entire population.

To demonstrate the first point let us take the first 7 natural numbers to represent the complete population. Such a population would not be anything like a cocked-hat distribution, in in fact there is no mode at all. If we took repeated random samples of 2 items from this population then by the laws of chance all alternatives should be chosen in equal proportions. As these 7 figures form an arithmetical progression, the first and last terms will give the mean. Thus $1+7$, $2+6$ and $3+5$ all average at 4, and form a mode at the average; in contrast $1+2$ at one end, and $6+7$ at the other, produce the averages of $1\frac{1}{2}$ and $6\frac{1}{2}$ only once, while an average of $2\frac{1}{2}$ can be formed two ways, by $1+4$, or $2+3$. If you work out all the possibilities you will see that you are getting a humped-back curve.

To demonstrate the second point, using the same numbers, you will see that where you take the means of the items two at a time, they will vary from $1\frac{1}{2}$ to $6\frac{1}{2}$. Three items at a time will range from $\frac{1+2+3}{3}$ to $\frac{5+6+7}{3}$, that is from 2 to 6; and if you take 5 items at a time the mean will vary only from 3 to 5. This relationship between the range given by the items themselves and the range given by the averages of different sets of items from the same population has been shown to be s.e. $= \frac{\text{s.d.}}{\sqrt{n}}$. That is, the standard error, indicating the range of the mean, is equal to the standard deviation of the sample, divided by the square root of the number of items in the sample.

Coming back to the question " Is there a real difference between our mean of 99·92° with an s.d. of 0·6856, and the text-book figure of 100·5°? " The standard error of our sample of 37 items will be, s.e. $= \frac{0{\cdot}6856}{\sqrt{37}} = 0{\cdot}1127°$, and this would be conventionally recorded as 99·92° F$\mp$0·113. As the sample contains over 30 items it will be sufficiently reliable for us to predict that if we took several other sets of samples, each of 37 items, and all from the same population, we would expect 2 out of 3 of the averages to fall within the range of 1 standard error each side of the average of the averages, which, itself, would fall in the middle of the distribution. Similarly we would expect 95 out of 100 of these averages to fall within a limit of 2 s.e. from the central average. Finally, we would expect only about 1 in every 400 averages to fall outside the range set by 3 s.e. each side of the centre average. Obviously this central average would be very close indeed to the true average for the whole population.

If, therefore, we can use the s.e. of our sample as a reliable yardstick then, even if the average for our sample is unusually high, or low, there should, nevertheless, be only 1 chance in 400 that is so high or low that it will be more than 3 s.e. away from the mean of the whole population. Our s.e. was 0·1127, so that 99·92$\mp$(3$\times$0·1127) gives a range of from 99·581 to 100·258° F. The text-book figure of 100·5° F. is outside this range so that we can judge it to be either the average of a different population, or else a pretty poor estimate of the true average of our population, based on either a very poor, or a very small, sample. Thus, although we cannot calculate any exact odds, because we know neither the number of animals sampled, nor the s.d. of the text-book range, we have shown a discrepancy that might be worth further examination.

Note that the standard error does, in fact, have some connection with the measurement of error. For example, if you take duplicate readings of a measurement these two readings will have an s.d., even if you do not work it out. The standard error will give you an indication of the range that the unknown correct reading is likely to be away from your actual readings. With duplicate readings this range is $\frac{\text{s.d.}}{\sqrt{2}}$. If you took 4 readings there is no reason why the s.d. should be larger, it will certainly be more reliable but it may even be smaller than with 2 items, whereas the s.e. will always be smaller, for it will be $\frac{\text{s.d.}}{\sqrt{4}}$, that is, only half the size it would have been if you had been content with only one reading. Thus to halve the error one must carry out 4 times the number of replicates.

Note that if we are recording the statistics of a cocked-hat curve for other people's benefit, we must not only record the standard deviation but also make it clear which type it is, that of the range of the items or that of the range of the mean, the latter being what we have called the standard error. Note also that, unless you mention the number of items, the reader cannot change one s.d. into the other, nor will he know how reliable your sample is. If you do not mention the number of items he can only guess that you are ashamed to mention it because they are so few, for not many people omit to mention the number if it is adequate.

Technical error

So far we have looked at a sample of temperatures and, because it gave a roughly cocked-hat distribution, we were able to use the standard deviation as a standard by which we could judge any horse with a temperature less than 98·0°, or more than 101·8° F. as abnormal. Further, we could use the standard error to predict that, had we the time to sample every healthy adult horse of that breed in that environment, the chances are 95 in 100 that this mean would fall within the range of $99{\cdot}92 \mp (2 \times 0{\cdot}1127)$, that is, roughly between 99·7 and 100·15° F. There is another advantage in dealing with a cocked-hat curve, and that is that a proper allowance can be made for chance errors that themselves fall in a cocked-hat distribution.

Even after some years' experience in a laboratory there are some people who remain quite blind to the existence of technical error, and are willing to accept any change in measurement as a change in the constitution of the experimental animal without considering, and certainly not measuring, the variation in estimations due to error alone. It might be thought that technical error in such a simple operation as taking a body temperature would be trivial but this is not so, and experiments have been ruined when the slight rises in temperature, indicative of the successful infection with, say, a virus, have been lost through a change in technical assistants at week-ends, or at other holidays.

The sample of temperatures we have been dealing with were gathered for clinical work, and here it is important to have as a standard a range that contains not only the physiological variation but the error of the method for, without the inclusion of the error, healthy animals would from time to time be regarded as diseased.

On the other hand if it were necessary to try to find out the limit for physiological variation alone an effort could have been made to cut down the error to such a small proportion that it would be negligible.

If we wished to try this method we could take the following steps to reduce the error: remove any hard faeces from the rectum, use the same accurate thermometer for each horse, mark the length of the thermometer to be inserted by means of a rubber ring, measure the time of insertion with care, record the average of two estimations, and use the pulse rate to check that the animal was not unduly excited.

Another source of error would be in reading the thermometer, for the clinical thermometer is calibrated in units of 0·2° F. and the mercury is read correct to the nearest mark. In trying to make a more accurate reading we might be tempted to sub-divide these graduations by eye alone, but this should be resisted for the impression of greater accuracy is mainly an illusion owing to the observor having "favourite figures." Yule and Kendall (1946, Table 6.4) tabulate such optical estimations and their example shows an obvious bias for readings of 0, 0·2, 0·5, 0·8, and 0·9; it can be seen that some observers tended to pile up results at each end of the interval, except that 0·1 seemed very unpopular. To read the thermometer accurately it would therefore be necessary to use some device, such as a micrometer eye-piece, so that the final digit could be read objectively against a visible scale.

Addition and subtraction of sums of squares

These temperatures were collected for clinical use without any special refinement in technique. If we now wish to find what part of this range was due to physiological variation alone, is there any way by which we could still obtain some estimate of this variation? Yes, it would be possible to obtain an objective estimate by taking advantage of the fact that sums of squares, or variances, of cocked-hat curves can, under certain conditions, be added or subtracted. Thus, if we find the sum of squares due to error alone, we can subtract this from that of the clinical range, and be left with the smaller sum of squares due to physiological variation alone. Before attempting this, however, let me try to demonstrate the principle on which it will be based.

To demonstrate the basis of this addition, or subtraction, the working model of Pearson's wedges can again be used, and the results are indicated in Fig. 71. To keep things simple

									SUMS OF SQUARES	THEORETICAL VARIANCE
				32/9					0	0
			16/8		16/10				32	1
		8/7		16/9		8/11			64	2
	4/6		12/8		12/10		4/12		96	3
2/5		8/7		12/9		8/11		2/13	128	4

Fig. 71. The simple addition of sums of squares.

we assume that an item deflected to one side of the wedge gains a unit of value, and deflected to the other side loses a unit of value. We start with 32 items each having the value of 9, thus 9 is the mean, and remains so down all the lines of wedges.

The first wedge deflects 16 items to the left and these take the value of 8, and the other 16 deflect to the right and take the value of 10. The sum of squares is found in the same way that was used in finding the s.d. so we now have:

Value	Frequency (f)	Deviation (d)	fd²
8	16	−1	16
9	0	0	0
10	16	1	16
	Total sum of squares		32

The sum of squares is therefore 32. In this example we are dealing with a theoretically perfect distribution so that to find the variance we divide by n, and not by $(n-1)$, because we do not need to guard against imperfect samples. The sum of squares is therefore divided by the total number of items, giving $\frac{32}{32}$, or a variance of 1.

The next row of 2 wedges will deflect the items farther to the extreme or back to the average, giving the following frequencies:

Value	f.	d.	fd²
7	8	−2	32
9	16	0	0
11	8	2	32
Sum of squares			64

Here the sum of squares is 64, and the theoretical variance will be $\frac{64}{32}$, or 2.

In Fig. 71 this process is continued to show the effect of 4 rows of wedges and it is suggested that in order to understand what is happening the reader might add the 5th row and calculate the sum of squares and variance for it.

Now we come to the point of the demonstration. From Fig. 71 we can see that if technical error was equal to 2 rows of wedges it would produce a sum of squares of 64 for 32 items, giving a variance of 2. If the physiological variation was equivalent to 3 rows of wedges it would produce a s.sq. of 96 for 32 items, and the variance would be 3. Our clinical survey of temperatures would include both these causes of variation and therefore be equivalent to 5 rows of wedges, with a s.sq. of 160 for 32 items, or a variance of 5. It can be seen that this amounts to the same figure as the two separate ranges added together. Thus:

	S. sq.	Variance
Error	64	2
Physiological	96	3
	—	—
Total Clinical	160	5

The model in Fig. 71 is very crude, for in actuality the difference made by each wedge would be more like 0·001°. Further, the technical error would represent about 40 lines of wedges and the physiological variation perhaps some 2000 rows. Nevertheless, the same principle would apply and, provided the number of items were the same, the sums of squares could be added; alternatively, if the number of items were different it would still be possible to add the variances.

To illustrate this last point, and to get a little nearer to the solution of our problem, let us take another theoretical example exhibited in Table 74. In this table we have a crude cocked-hat curve consisting of 4 horses with a body temperature of 99·0°, 8 at 100° and 4 at 101°, but in taking these temperatures we suppose that 4 thermometers were used. Suppose these thermometers were tested later on in the laboratory and registered 99·9, 100·0, 100·0, and 100·1 for a known temperature of 100·0. At the top of the table the items are so arranged that every grade of temperature is estimated by every thermometer, which gives us an easy theoretical equivalent of the thermometers being used equally over a large random sample. This sample of 16 horses, with the same temperature distribution as the four horses, gives a theoretical variance of 0·505. It can be seen that when we subtract the thermometer error variance we have $0{\cdot}505-0{\cdot}005=0{\cdot}500$.

TABLE 74

COMBINED ERROR AND NATURAL VARIATION

Natural Variation in Horses' Temp.	Error in Thermometers	Combination of Error and Natural Variation	d	d^2
99·0	−0·1	98·9	−1·1	1·21
99·0	0	99·0	−1·0	1·00
99·0	0	99·0	−1·0	1·00
99·0	0·1	99·1	−0·9	0·81
100·0	−0·1	99·9	−0·1	0·01
100·0	0	100·0	0·0	0·00
100·0	0	100·0	0·0	0·00
100·0	0·1	100·1	0·1	0·01
100·0	−0·1	99·9	−0·1	0·01
100·0	0	100·0	0·0	0·00
100·0	0	100·0	0·0	0·00
100·0	0·1	100·1	0·1	0·01
101·0	−0·1	100·9	0·9	0·81
101·0	0	101·0	1·0	1·00
101·0	0	101·0	1·0	1·00
101·0	0·1	101·1	1·1	1·21
		16) 1600·0	0·0	8·08
		100·0		

Sum of squares : 8·08

Theoretical Variance : $\frac{8 \cdot 08}{16} = 0 \cdot 505$

Standard Deviation : $\sqrt{0 \cdot 505} = 0 \cdot 71063$

SEPARATE ERROR AND NATURAL VARIATION

Variation in Thermometers

Thermometer Error	d	d^2
99·9	−0·1	0·01
100·0	0·0	0·00
100·0	0·0	0·00
100·1	0·1	0·01
	0·0	0·02

Sum of squares : 0·02

Theoretical variance : $\frac{0 \cdot 02}{4} = 0 \cdot 005$

Natural Variation in Temp.

Temp.	Freq.	d	fd	fd^2
99	4	−1	−4	4
100	8	0	0	0
101	4	1	4	4
			0	8

Sum of squares : 8

Theoretical variance : $\frac{8}{16} = 0 \cdot 5$

Addition of variances : $0 \cdot 005 + 0 \cdot 5 = 0 \cdot 505$.

This method need not be limited to the errors in the thermometers, other errors could be measured. Thus each worker concerned could be asked to read, say, 30 different temperatures to measure the error in reading thermometers. This error could also be subtracted, again slightly narrowing the range. The practicability of this method varies with the type of estimation, but if you are interested in the method for your own work you might wish to consult the paper by Berkson, Magath and Hurn (1940). These authors break the total error in counting erythrocytes into the error of distribution within the counting chamber, the error between one chamber and another, and the dilution errors given by different pipettes.

Estimation of error by duplicate measurements

The clinical survey on temperatures does not lend itself to an easy analysis of errors for much of the error lies in the distance that the thermometer is inserted into the rectum, and this difficulty varies from animal to animal. But with the knowledge that we can add or subtract variances, we can estimate the total error by taking duplicate measurements. We have sampled 37 horses and found a mean of 99·92° F. and an s.d. of 0·6856°. Now the question is what part of the range is due to the physiological variation in the horses and what part is due to the error of the method. Does technical error represent half the range? Or is it just a small proportion? Obviously with biological material we are not going to get an answer that is mathematically perfect, but we can get one that is both objective and sufficiently accurate to have a practical value.

The clinical survey was shared by 2 workers, *A* and *B*. And between them they used 3 thermometers; the actual thermometer used for each horse depending on the one that came to hand. The test is shown in Table 76 and it can be seen that it contains a comparison of each thermometer with the others, and comparisons between each worker with himself and with his companion.

The test consists of duplicate readings on 10 horses that chanced to be at the laboratory and these did not necessarily belong to the same population as the 37. The whole test took well under one hour and therefore permitted little time for physiological variation in the individual animals. The most likely physiological variation would be that due to excitement, which would raise the temperature. As the second reading shows only two increases there is no reason to believe that excitement interfered with the test. It can be seen that *B* in the 4 comparisons with *A* obtained a higher figure in 3 instances. This could easily be due to chance but it might have been due to the fact that in using so-called half-minute thermometers, *A* kept to his habit of leaving it inserted for 45 seconds and *B* to his habit of inserting it for 60 seconds. Here, the cause of the discrepancy is quite unimportant for the object is only to reproduce and measure the technical errors, not to cut them out.

TABLE 76

Duplicate Body Temperatures as measured by 3 thermometers and 2 attendants as a measurement of technical error.

Horse	Therm. No.	Worker	Temp.	Therm. No.	Worker	Temp.	d.	d²
1	1	A	99·6	1	A	99·6	0·0	0·00
2	2	B	100·6	2	A	100·6	0·0	0·00
3	3	A	100·6	3	A	100·2	−0·4	0·16
4	1	B	100·0	2	A	99·8	−0·2	0·04
5	2	A	99·2	3	A	99·2	0·0	0·00
6	3	B	100·8	1	B	100·8	0·0	0·00
7	1	A	99·4	3	B	99·8	+0·4	0·16
8	2	B	100·6	1	B	100·4	−0·2	0·04
9	3	A	99·4	2	B	99·6	+0·2	0·04
10	1	B	100·6	3	B	100·6	0·0	0·00
				Sum of squares of differences				0·44

$$\text{Sum of squares } \frac{0{\cdot}44}{2} = 0{\cdot}22$$

$$\text{Variance } \frac{0{\cdot}22}{10} = 0{\cdot}022$$

		Variance
Clinical Survey (s.d 0·6856)²	=	0·4700
Technical Error		0·0220
Physiological variation in horse		0·4480

S.d. of physiological variation $= \sqrt{0{\cdot}448}$
$= 0{\cdot}6693°$ F.

We use these duplicate readings in the following way. If we look at those for horse 3 we see that the first estimate was 100·6 and the second 100·2. The best estimate for horse 3 is therefore $\frac{100{\cdot}6+100{\cdot}2}{2}$, or 100·4. The deviations from this mean will be (100·4—100·6) and (100·4—100·2) and the deviations squared will be $(-0{\cdot}2)^2+(0{\cdot}2)^2$, giving 0·08. We can, however, find this in a quicker way by squaring the difference in the duplicates and then dividing by 2. Thus we have $(100{\cdot}6-100{\cdot}2)^2 = 0{\cdot}16$, which divided by 2 gives 0·08, as before.

Using this second method we find the sum of the squares of these differences is 0·44, and this divided by 2 gives the sum of squares of 0·22. Each duplicate provides one independent comparison, so that there are 10 degrees of freedom, which gives a variance of 0·022° F. At the bottom of the table we see that this variance can be subtracted from the clinical survey variance to leave that part of the range that we now estimate is due to physiological error alone.

When we inspect the figures we see that all we have done is to reduce the s.d. of 0·6856 to 0·6693. This is a trivial reduction but the important point is that, having made an objective estimate of the error, we can use, and defend, our range with complete confidence.

Variance ratio

From what we already know it is obvious that if 100 shot are poured down Pearson's wedges the results are not exactly the same each time although, as the wedges are the same, the trials can be looked upon as samples from the same population. Further, if the cups into which the shot fall had ascending values, say, from left to right, then the variance for each 100 shot would be similar but not identical. Out of every two variances, therefore, one could put the larger on top and divide by the smaller to get an index figure of over 1, and sometimes perhaps as large as 3 or 4. If this was continued long enough a table could be built up showing what size indices were common and what size occurred as seldom as once in 20 times. Varying the number of shot in each trial, and finding the indices for groups of variances, rather than just pairs, would provide enough work for even the most energetic do-it-yourself enthusiast. It is therefore very fortunate that similar, or better, results can be obtained by using mathematical formulae, and such results are recorded in what Snedecor, in honour of Fisher, has called the F Table.

The main use of this table is for the analysis of variance, which will be introduced in the next chapter, but at present its use will be restricted to giving the probabilities for a direct comparison of variance. A more complete table is given by Fisher and Yates as Table V, Variance Ratio.

To see why it is useful to compare two variances we can suppose that we have two groups of animals, and that one group of them is badly infested with intestinal worms. After measuring the haemoglobin values in both groups we find the average value is very much the same in both groups. If we have drawn rough histograms to show the distributions we may notice a wider distribution in the affected group, for the worm infestation may have caused anaemia in some animals and diarrhoea in others. Severe diarrhoea may cause a loss of water from the blood giving an anhydraemia, so that the haemoglobin values are quite high. As the low figures for anaemia, and the high figures for anhydraemia tend to compensate each other, the mean is not greatly affected. With the means the same value, the only way of showing the two groups do not belong to the same population is by comparing the variances and looking up the index in the table of F of which a sample page is shown in the appendix.

To show how the test is applied, assume that there are 13 healthy animals with an average haemoglobin value of 12·0 g./100ml. and a variance of 1·640, and that the 9 infected animals average 11·6 g. and have a variance of 9·627. To use the table we divide the larger variance by the smaller one, giving

$$\frac{9{\cdot}627}{1{\cdot}640} = F = 5{\cdot}87.$$

The group with the larger variance will be n_1, and for the 9 items the degrees of freedom will be 8. The smaller variance will refer to n_2, and as 1·640 was for 13 items the d.f. will be 12. Thus we will look along the headings of the columns for the figure 8, and down the side for the row opposite 12. In Fisher and Yates' tables there are different pages for the variance ratio, or F., for probabilities of 20 per cent, 5 per cent, 1 per cent and 0·1 per cent. But in this problem, by always putting the larger variance on top, we are using $\frac{9{\cdot}627}{1{\cdot}640}$ to represent F values of both $\frac{9.627}{1{\cdot}640}$ and $\frac{1{\cdot}640}{9{\cdot}627}$, so that we are using one tail of the probability curve to do the work of both. Thus to obtain the conventional double-sided probability we must double the recorded probabilities.

Because of this the 5 per cent page represents 10 per cent, and therefore, when with $n_1 = 8$ and $n_2 = 12$, we find $F = 2{\cdot}85$ then, although our result is higher than this we cannot claim significance. We must therefore go farther and look up the 1 per cent page, which for our purpose represents 1 in 50. On this page we find 4·50, and, as our result of 5·87 is greater than this, we can claim significance. Thus we can state that it would be unlikely for the two lots of items to have been distributed by the same set of wedges, or to be more orthodox, to be random samples from the same population.

When dealing with the mean and its standard error we found that we could use these statistics to predict the range over which the true mean of the population might lie. Can we also predict the limits for the true standard deviation of the population of which we have only a sample? Yes, in the same way that the s.e. of the mean is shown by $\frac{\text{s.d.}}{\sqrt{n}}$ for large samples the standard error of the standard deviation is calculated from $\frac{\text{s.d.}}{\sqrt{2n}}$. Small samples, say under 50, require a more complicated formula involving χ^2, and this is given by Moroney. But in routine experimental work it is unlikely that you will require to use either of these formulae.

Summary

Where a sample from a population has been shown to be distributed as a cocked-hat curve we can calculate various statistics that can be used for prediction.

In the last chapter it was seen that the standard deviation could be used to predict how likely it was for fresh items to fall between different limits of measurement. In this chapter we see that by using the standard error we can predict the limits within which the mean of the whole population, of which we have only a sample, is likely to lie.

Further, where all the distributions concerned are humped-backed, we can add or subtract variances. Thus by calculating the variance of technical error, and that for error plus natural variation in the material, we can subtract the one from the other and predict that the remainder represents the natural variation in the material.

Reference

Berkson, J., Magath, T. B. and Hurn, Margaret (1940). "The Error of Estimate of the Blood Cell Count as made with the Haemocytometer." *Amer. J. Physiol.* **128**, 309.

Chapter 5

ANALYSIS OF VARIANCE

At the present time any novice in the theory of estimation should be able to set out the calculations necessary for making estimates, almost, if not quite, as good as they can possibly be.

R. A. Fisher

We can now use what we have learnt about the addition of sums of squares to deal with a very common problem. That is, to decide if two sets of samples are likely to have come from the same population, or whether the items in one group have been affected by some additional factor and must now be regarded as belonging to a different population. But before demonstrating the orthodox, sensitive, method that involves using sums of squares, it will be shown that an easy makeshift method, using the binomial distribution, may sometimes be sufficient to demonstrate significance.

Comparing two sets of measurements

As an example in comparing 2 sets of measurements let us imagine that a flock of lambs was too numerous to be grazed in a single field. Because of this the flock has been split roughly into halves and each half has been placed in a different field. After a few weeks it is noticed that the lambs in one field are not thriving as well as those in the other, and immediately a stream of hypotheses pass through the mind.

Before paying any serious attention to these hypotheses the first action of a scientist will be to confirm the impression that one part of the flock is thriving less than the other, and to do this the lambs would be weighed.

Let us assume that a frequency table for the lambs was as follows:

Weight of Lambs in Lb.

Veight:	94	96	98	100	102	104	106	108	110	112	114	116	118	120	122	124	126	128	Total
oor Gp.	1	1	1	1	.	1	1	2	1	2	.	1	1	.	1				14
Iealthy Gp.				1	.	1	1	.	1	1	2	2	2	1	.	1	2	1	16

From this frequency table it seems likely that what caught the eye was the 4 lambs of 100 Lb. or less in the poor group, in contrast to the 4 lambs weighing over 122 Lb. in the other group. It can be seen that each group has some tendency to cluster near the middle of the range, the poor group between 108 and 112 Lb. and the healthy group between 114 and 118 Lb., and this is sufficient indication that these samples are from cocked-hat curves for us to use methods based on the normal distribution.

Use of binomial tables

As the binomial distribution is related to the normal distribution, and as we have already mentioned some tests and tables that can be applied to enumeration data, we might begin the easy way by seeing if we can use this knowledge to show that the two groups are heterogeneous. On the left of the frequency table it can be seen that we can cut off 4 of the low weights—those under 102 Lb. —with only 1 of the healthy group. We can therefore ask the question: On the null hypothesis, that those 30 items all come from the same binomial distribution, what is the chance of finding 1 *A* in a group of 16 in 1 sample and 4 *A*s in a group of 14 in a second sample?

Looking up Table IV given by Mainland, Herrera and Sutcliffe, we find under $N_1 = 16$, $N_2 = 14$, $p =$ Less than 0·025 that the first entry with a unit value is 1/7; this indicates that less extreme proportions, such as 1/6, or our 1/4, are not significant. This test has failed so we might try something with a bigger weight of figures in each cell. A good division might be to take items above and below 113 Lb. This gives the following 2×2 table:

	<113	>113 Lb.	
Poor	11	3	14
Healthy	5	11	16

In Table IV we find N_1 16, N_2 14. gives 5/11 (0·0121). This signifies that a proportion of 5 *A*s in the 16 group to 11 *A*s in the 14 group would occur by chance 1·21 times in 100. The conventional double-sided significance would be twice this, that is, $2 \times 1{\cdot}21$, or 2·42 times in 100, and the answer is significant. We can say, therefore, that if the samples come from something similar to a binomial distribution, and a wedge was arranged to throw the items into categories above, or below, 113 Lb. one would not expect such contrasting trials to be due to chance alone.

As the result was significant there is a real reason for seeking an hypothesis and taking action on it; alternatively one could go on to the more sensitive orthodox test knowing that this would be likely to increase the significance. On the other hand, had the result been non-significant, only a few minutes would be wasted and, if the result was anywhere near significance, one could still go on to the more sensitive test in the hope that this result would be within the border of significance.

Analysis of variance

We come now to the orthodox, and most sensitive, test for determining if these two groups are likely to be from the same population, and for this purpose the two groups are detailed once again in Table 85. Inspecting the weights we see that the arithmetic can be eased by coding the items by removing the value of 100 from each item. Thus the lowest value, 94, becomes −6, and the highest, 128, becomes 28. The rest of the calculation is carried out on these coded figures.

From Table 85 we see that the coded items for the healthy group total 252 and give an average of 15·7500, and the poor group total 104 with an average of 7·42857. The average for the whole flock is $\frac{252+104}{30}$, or 11·86. To test the likelihood of there being a real difference between these two groups we make the assumption that if we could remove the error due to the natural variation in lambs of that age and breed, and the variation due to error in weighing, then all the lambs in the healthy group would weigh 15·75 Lb. and those in the poor group would weigh 7·42857 Lb. Therefore the deviations of these lambs, considered in groups, compared with the average for the whole flock would be 16 $(15{\cdot}75-11{\cdot}8\dot{6})^2$ for the healthy group and 14 $(7{\cdot}4286-11{\cdot}8\dot{6})^2$ for the poor group. It can be seen under Sums of Squares Between Groups in Table 85 that these squared deviations amount to 517·03. Because in most experiments one group is treated in one way and another is treated in a different way, this sum is usually spoken of as the sum of squares between treatments.

The next step is to find the range for the whole flock, that is, the difference between each item and the mean for the 30 items. One of the short cuts demonstrated in Chapter 3, was that to avoid fractions you could use a convenient false mean for finding a false sum of squares and then use a correction term to remove the error. This correction term is $\frac{(\text{error})^2}{\text{no. of items}}$. Thus if you have items 2, 3, 4, and 5, you could avoid using the awkward mean of 3·5 by choosing either 3 or 4. If you subtracted 4 from each item the deviations would be -2, -1, 0, $+1$, giving an error of -2. The deviations squared would come to $4+1+1 = 6$, and the correction term $\frac{(-2)^2}{4}$, or 1. Giving the sum of squares $= 6-1 = 5$. Workers who carry out analyses of variance usually have some type of calculating machine, and they can save time by choosing a false mean of zero. By this method 2, 3, 4, and 5 give an error equal to their total, and the deviations equal the original items squared, that is $4+9+16+25$ making a false sum of squares of 54. The correction term will be $\frac{(14)^2}{4}$, or 49, and the true sum of squares will be $54-49$, or 5, as before.

In Table 85, therefore, we find the total s.sq. by squaring all the items and subtracting the Main Correction Term. The False sum of squares is 6648 and the correction term is $\frac{(356)^2}{30}$, leaving a true sum of squares of 2423·5. As a digression, note that $\frac{(356)^2}{30}$ is the same as the total, 356, multiplied by the mean, $\frac{356}{30}$. Also, if you are interested, ask yourself if there is any difference in the figures we have used, and those that would have been obtained if we had left the data uncoded, and used a false mean of 100 Lb.

TABLE 85

Comparison of the Weights of 2 Groups of Lambs in Different Fields.

Field 1 (? Healthy)			Field 2 (? Poor)		
Wt. Lb.	Coded Items	(Items)^2	Wt. Lb.	Coded Items	(Items)^2
100	0	0	94	−6	36
104	4	16	96	−4	16
106	6	36	98	−2	4
110	10	100	100	0	0
112	12	144	104	4	16
114	14	196	106	6	36
114	14	196	108	8	64
116	16	256	108	8	64
116	16	256	110	10	100
118	18	324	112	12	144
118	18	324	112	12	144
120	20	400	116	16	256
124	24	576	118	18	324
126	26	676	122	22	484
126	26	676			
128	28	784		104	1688
	252	4960			
Average 15·75			Average 7·42857		

Data for Entire Flock

No. of Items	Value of Coded Items	Sum of (Items)^2
16	252	4960
14	104	1688
30	356	6648

Total Sum of Squares

Sum of Items squared $= 6648$

Main C.T. $= \frac{356^2}{30} = 4224 \cdot 5\dot{3}$

$2423 \cdot 5$

$$\text{Average} = \frac{356}{30} = 11 \cdot 8\dot{6}$$

Sum of Squares Between Groups.

Healthy Group $16\,(15 \cdot 75 - 11 \cdot 8\dot{6})^2 = 241 \cdot 28$

Poor Group $14\,(7 \cdot 4286 - 11 \cdot 8\dot{6})^2 = 275 \cdot 75$

$517 \cdot 03$

ANALYSIS OF VARIANCE

	Sums of Squares	Degrees of Freedom	Variance	F.
Total	2423·5	29		
Between Groups	517·0	1	517·0	7·59
"Error"	1906·5	28	68·1	

We have now got the sum of squares for a difference in the 2 groups, which we are presuming is due to some factor or treatment affecting 1 group only. We have also the sum of squares for the whole range in weights. Under the heading Analysis of Variance in the table, it will be seen that we subtract the part from the whole, on the assumption that this difference will represent that part of the range due to physiological variation in lambs of that age and breed, and the error of weighing. As in the example of body temperatures we will expect both of these variations to give cocked-hat distributions.

We are dealing with 30 items which will provide 29 degrees of freedom. We may look upon the difference in the two groups as due to the action of a single wedge that deflects items to an average of 15·75 lb. on one side, and 7·42857 on the other; the difference in treatment therefore provides 1 degree of freedom. This leaves 28 d.f. for the wedges concerned with error. We are going to use this information by comparing the size of the mean square, or variance, so that we divide the sums of squares by the degrees of freedom to produce variances of 517·0 and 68·1.

Our null hypothesis is that all the items belong to the same population, which we have previously looked upon as equivalent to coming down the same set of wedges. The question now is: If the average variance for groups of shot coming down the wedges is 68·1 how likely will it be for the variance between two sets of shot to be as high as 517, or 7·59 times as high?

Using Fisher and Yates' Table V for Variance Ratio, we take the larger variance as n_1. In this example this refers to the degrees of freedom between groups and is 1; n_2 refers to the degrees of freedom for " error " and is 28. The first page of the table is for 20 per cent probability, here the intersection gives 1·72. Our result is higher than this so that we can move to less likely probabilities. The next page is for a probability of 5 per cent ($p = 0{\cdot}05$). Here 1 and 28 intersect at 4·20; our result is higher than this and is therefore significant. On the next page for 1 per cent we find 7·64; our result of 7·59 is nearly at this figure, so that for practical purposes we can say that $p \simeq 0{\cdot}01$. Thus the result is very near the high significance level and the difference between the two groups is most unlikely to be due to chance alone.

Had we used Snedecor's F Table, we would have taken f_1 as 1, and f_2 as 28 and at the point of intersection we would have found 4·20 and 7·64, denoting significance and high significance.

SIGNIFICANCE AND ACTION

Before leaving this example let us look more closely at what is at stake. A flock of lambs has been divided into two fields. When first split the lambs in each field appeared to be of roughly equal weight, but after a few weeks one group carried more weight than the other to give the comparative averages of 115·75 and 107·43 Lb. A binomial test gave $p = 0{\cdot}0242$, estimating that such heterogeneity would occur only 2 or 3 times in 100. On this basis, if the items really were shot falling over wedges, then, on the average, one would make a mistaken decision 2 or 3 times in 100. But we are not dealing with shot we are dealing with biological material. Biological measurements tend to cluster round the mean and extreme measurements are likely to be pruned off in the course of Natural Selection by the resulting inefficiency or disease. For this reason even the conventional $p = 0{\cdot}05$, for significance, is remarkably reliable and I have neither heard, nor read, of an instance where, provided the logic was correct, the prediction given by this probability has subsequently been shown to be wrong.

If we therefore accept as fact that there is a real difference in the two groups, what action must be taken? This depends on circumstances. Depending on the hypothesis formulated to explain the loss of weight one might feed a supplementary ration, dose the lambs for intestinal worms, or move the lambs into another field. On the other hand, if the lighter lambs appeared healthy, and the difference of 10 per cent in average weight between the 2 groups was not commercially, or experimentally, important, it might be decided to take no action at all except to watch that the difference did not become too extreme.

We have described the consequences following a significant result from a binomial test. What now would be the consequences of showing a highly significant result from the analysis of variance? The answer is: No difference! The need for action does not depend on the degree of significance, it depends on the amount of weight lost. The urgency of action would be greatly increased if weighing had shown that there was a 20 per cent loss in the lighter group, even if the groups were so small that it was hard to prove significance. Significance alone does not prove the need for action, thus it is that one can prove by taking large samples that the greater height of the Scotsman, compared to an Englishman, is very highly significant, nevertheless, as yet, nothing has been done about it. Significance is important in judging if an apparent change is likely to be due to chance alone; the degree of significance is important in predicting if the same treatment will produce the same result in the future, but this is discussed in Chapter 7 under the heading of Significance and Faith.

Paired data

Very often in simple experiments we obtain data in which the results are arranged in pairs, such as the familiar " Before " and " After " treatment of advertisements. As an example let us suppose that we are interested in 4 mature guinea pigs, and we wish to know if, with the diet they are receiving and the environment in which they are kept, they are still gaining in weight. To determine this we have weighed the guinea pigs and then reweighed them at the end of 14 days, obtaining the following results:

Table 88

Weight in Grams of 4 Mature Guinea Pigs, Weighed Twice With an Interval of 14 Days

G.P.	1st Wgh.	2nd Wgh.
A	704	708
B	717	720
C	725	731
D	712	719

To inspect these figures easily we will code them by subtracting 700 g. from each weight, and tidy them by putting the first weights into an array. This gives us:

Table 88 Coded

G.P.	1st		2nd			Gain in weight
A	4		8			Yes
D	12		19			Yes
B	17		20			Yes
C	25		31			Yes
	—		—			
Totals	58	+	78	=	136	
Av.s	14·5		19·5		17	

Looking for the easiest way to show a significant difference we might try a simple binomial method. This would be based on the null hypothesis that there had been no true gain in weight but that the result was due to the play of a large number of chance factors, such as the animals eating more, or less, than usual, drinking more, or less, than usual, or sweating more, or less, than usual. By the laws of chance these multiple factors that can increase, or decrease, weight would, on the average, produce an increase in weight in 2 of the animals and a decrease in the other 2. This expectancy is therefore equivalent to the results for tossing a coin. As the actual result has shown that all 4 animals gained in weight this would be equivalent to throwing a coin 4 times and getting the same result each time. The chance of getting 4 successive " heads " is 1 in 16, and of getting " tails " is the same, so that the conventional double-sided probability is 2 in 16, or 1 in 8, which is not significant.

This test, in which the null hypothesis provides a theoretical value of $\frac{1}{2}$ — and $\frac{1}{2}$ + is known as the " sign test." It cannot provide a significant answer to our problem because there are too few pairs of observations and the evidence provided by the size and regularity of the gain is ignored.

We have failed to show significance by the sign test, can we test significance comparing the two sets of weights as we did with the lambs? No, we cannot use a simple comparison of one group with the other for both groups contain the same 4 guinea-pigs, and therefore we are dealing with multiple samples of the same animals. Thus if we were making a histogram we could use the 4 averages but we would not be entitled to use the 8 weights as 8 bricks in the histogram, nor divide by 8—1 when finding the standard deviation.

Because of this the correct test is the *t* test on paired data which we will carry out as an analysis of variance. But before doing so let us carry out the test as though we had forgotten to mark the pigs and thus could not pair them. From this test we shall be able to see that although we will divide the error sum of squares, or residue, by 8—1 instead of 4—1, the test will be unable to detect any significance.

To do this we can go back to the figures in Table 88 Coded. As in the test on lambs we can use a false mean of zero so that the true total sum of squares will be equal to the sum of all the items squared minus the main correction term. The sum of the items squared is $(4^2+12^2..20^2+31^2) = 2860$, and the main correction term is $\frac{T^2}{n}$, or $T\times\frac{T}{n}$ (Total times average) giving $\frac{136^2}{8} = 2312$. So that the total sum of squares is $2860-2312 = 548$.

Again we can find the sum of squares for the difference in columns, or treatments, by removing all the apparent error and giving each item in that column the average value for the column. With these guinea pig weights we would thus have the following items: 14·5, 14·5, 14·5, 14·5, 19·5. 19·5, 19·5, 19·5, with an average of 17. The difference of these items from the average would be 8(2·5) and the sum of the squares of these deviations would be 8(6·25), or 50.

We can subtract this 50 from the total of 548 and get the figure of 498 for the sum of squares for all the other sources of variation, known or unknown.

For 8 items we have 7 degrees of freedom and this gives us the following analysis of variance:

Source	Sum of squares	d.f.	Variance	F
Total	548	7		$\frac{50}{83} = 0{\cdot}6$
Between Columns	50	1	50	
	—	—		$\frac{83}{50} = 1{\cdot}66$
" Error "	498	6	83	

It can be seen that the conventional F value obtained by dividing the " between column " variance by the " error " variance gives a value of less than 1 and is far from significant.

As a digression, note that the F value for $\frac{\text{error variance}}{\text{between col. variance}}$ is higher, and more than 1. Has this any significance? It is unusual for the variance for error to be greater than that for treatment, but let us see just how high this figure must be to indicate significance. Firstly we must note that the F table gives the single-sided significance for $\frac{\text{columns}}{\text{error}}$, for this is the logical requirement. In using the Table for $\frac{\text{error}}{\text{columns}}$ we are admitting that error can have the larger value, and therefore we require a double-sided significance. Because of this the Table for $p = 0{\cdot}05$ now represents $p = 0{\cdot}1$, which is not significant. We must therefore look up the table for $p = 0{\cdot}01$, which now represents $p = 0{\cdot}02$, or 1 in 50. Further, although with $F = \frac{\text{columns}}{\text{error}}$ n_1 in the table was 1 and n_2 equalled 6, now we are using $F = \frac{\text{error}}{\text{columns}}$, and $n_1 = 6$, and $n_2 = 1$. Looking this up in Fisher and Yates' Table V at the 1 per cent level we find the exceptionally high figure of 5859, so that there is not the slightest suggestion that our figure of 1·66 is in any way abnormal.

Routine simplification in finding sums of squares

Before going on to the correct method of testing these pairs of weights it is advisable to examine a more widely applicable method of finding sums of squares, for this method will be used in the calculation. The point that will be made in this section is that the sum of squares between columns is equal to the sum of the correction term for each column, minus the main correction term. In symbols this would be: S.sq. between columns = $C.T.(A)+C.T.(B)-M.C.T.$ If, however, you are willing to accept this as a fact there is no need to read the rest of this section and you can pass on to the actual test, coming back to this section later if interest prompts you.

Up till now we have used the knowledge that: Total s.sq.— s.sq. between columns = s.sq. for error. But this s.sq. for error, that is the s.sq. for all the other variations that exist within the groups, can be found in another way. That is, by finding the sum of squares of the deviations from the mean for each column separately, and then adding them together. Because of this it can be termed the sum of squares within the columns. To show this using our example we can use the following table:

Direct Calculation of the Sum of Squares Within Treatments

	A.	Items ²		*B.*	Items ²
	4	16		8	64
	12	144		19	361
	17	289		20	400
	25	625		31	961
Totals	58	1074		78	1786
Av.	14·5	841	Av.	19·5	1521
C.T.(A)	841·0	233	*C.T.(B)*	1521·0	265

$$233+265 = 498$$

From the table it can be seen that the figures used for the calculation of the s.sq. within Column *A* are as follows: Sum of items² = 1074, minus the correction term for *A*, $58 \times 14{\cdot}5$, giving 233. Similarly, for *B*, $1786-78 \times 19{\cdot}5 = 265$. Making the sum of squares within columns as $233+265 = 498$.

We can use this information to find the s.sq. between the columns because we already know that the total sum of squares is equal to the sum of squares between plus those within the columns. So we have:

Source	Sum of squares
Total	548
Within	498
Between	50

We found earlier in this chapter that the total sum of squares was given by the sum of all the items squared, 2860, minus the main correction term, 2312. So that rewriting the above small table in detail, using the figures from our example, we have the following:

Sum of squares between columns equals:

$$(2860-2312)-(1074-841+1786-1521)$$

Because 1074+1786 must equal 2860 these terms cancel out. So that if we correct for the double negatives in the second bracket, we have:

$$841+1521-2312 = 50$$

Or in symbols:

$$C.T.(A)+C.T.(B)-M.C.T. = \textit{S. sq. between columns.}$$

Which is the formula we will use in the succeeding test.

We have tried to show a significant difference between pairs of guinea pig weights, but we were unable to show that these gains indicated a true gain in weight, and not just an apparent one. Using a simple analysis of variance we found that the error, that is, the variation due to all factors other than those affecting one column at the expense of the other, was even larger than that between these two columns. When we look at the coded figures for the first weighing we see that a very high proportion of the total variation is due to the fact that the guinea pigs were of different weights at the first weighing. If instead of having these weights of 4, 12, 17 and 25 we had chosen 4 guinea pigs at exactly 14·5 then the whole sum of squares of 233 within the first column would be non-existent, and that within the second column would be much smaller. The smaller the sum of squares for error, the easier it would be to obtain a significant difference between the variances. What we require, therefore, is a method that will permit us to remove the sum of squares that is due to the natural variation in average weight in these guinea pigs, and this is done by making an additional subtraction in the analysis of variance. The calculations in this method is shown in Table 93.

TABLE 93

Analysis of Variance on Paired Data Demonstrating a Significant Increase in the Coded Weights of 4 Guinea Pigs Over the Period of a Fortnight.

Animal	1st Wt.		2nd Wt.		Sub. tot.	Av.	*C.T.*
A	4		8		12	6	72·0
D	12		19		31	15·5	480·5
B	17		20		37	18·5	684·5
C	25		31		56	28·0	1568·0
Sub. tot.	58	+	78	=	136		2805·0
Av.	14·5		19·5		17		
C.T.	841·0		1521·0		2312		

Sum of items ^{2} = 2860.

Calculation of Sums of Squares :

Total sum of squares	2860 − 2312 = 548
Sum of squares between columns	841 + 1521 − 2312 = 50
Sum of squares between rows	2805 − 2312 = 493

Analysis of Variance

Source	S. sq.	d.f.	*V.*	*F.*	
Total	548	7			
Between Columns	50	1	50	30·00	Sig.
Between Rows	493	3	164·3	98·60	Highly Sig.
Error	5	3	$1 \cdot \dot{6}$		

In the table we see that, as before, the sum of the squared items minus the main correction term gives a total s.sq. of 548. And the sum of the correction terms for the columns, minus the M.C.T. gives the s.sq. between columns. Now we come to the additional step and find the correction term for each row. The sum of these correction terms, minus the M.C.T. gives us the s.sq. for between rows, in the same way that it would if we turned the table round so that the rows would, indeed, be the columns.

For the analysis of variance from these figures we still have the 7 degrees of freedom for the 8 items, and 1 is still taken for the comparison of the two columns. Three wedges would be required to split the items into 4 rows so that out of the 6 remaining degrees 3 are taken by the rows, leaving only 3 for error.

The sum of squares for error is now the difference between 548 and 50+493, leaving only 5. The variances are $\frac{50}{1}$ for between columns, $\frac{493}{3}$ for rows, and $\frac{5}{3}$ for error. Thus the *F* value for the columns is now $\frac{50}{1 \cdot 66}$, or 30·00. Looking this up in Fisher and Yates' Table V with $n_1 = 1$, and n_2 now only 3, we find $p = 0 \cdot 05$ has a value of 10·13, and $p = 0 \cdot 01$ a value of 34·12. It can be seen that now we have removed the variation due to using animals of different weights, the gain in weight over a fortnight is significant, and nearly highly significant. Giving evidence that there has been a real gain in weight.

The *F* value for between rows is even higher, being 98·6. Seeking the probability when $n_1 = 3$ and $n_2 = 3$ we find the value for $p = 0 \cdot 01$ is 29·46, and for $p = 0 \cdot 001$ is 141·1. There is therefore a highly significant difference between rows with

$p = 0{\cdot}01/0{\cdot}001$. This fact is, however, of no importance to us because all it indicates is that with the error of weighing plus that of other chance increases, or decreases in weight, having a variance of only $1{\cdot}\dot{6}$ the differences in the average weights of these guinea pigs is unlikely to be due to error alone. More bluntly, the guinea pigs are definitely not all the same weight.

Paired data and false pairing

In this last problem, in which 4 guinea pigs were weighed twice, we have a demonstration of the importance of numbering animals, and recording the data under these numbers. If we had weighed the 4 animals in the same wire basket all we would have had for comparison would be one weight of 2,858 g. and another of 2,878 g. Here there would be nothing on which we could estimate likelihood.

By weighing each pig separately we were able to estimate the variation due to factors other than the difference between the two columns. But in this instance, although we obtained an estimate of error, the estimate was so big that the difference in columns could easily have been the result of chance.

By weighing each pig separately, and marking the pig so that we knew which weight belonged to which animal, we were able to go a step farther, and thus show that each individual pig had gained in weight, and the amount of weight it had gained. This allowed us to carry out an analysis of variance on paired data. In this analysis the sum of squares produced by the animals each being a different weight was also subtracted from the total sum of squares. This left a very small residue against which it was possible to show that the increase in weights in the 2 columns was significant.

Note that the difference in rows, as well as columns, was significant—showing that the different weights of the different pigs was not just an error in weighing. It was this significance in the differences between rows that allowed us to show the significant difference between the columns. In fact, in other circumstances, where there was no significant difference between individual animals, it might have been better not to extract this sum of squares, for then the residual sum of squares would have been divided by 6, as with the test on unpaired data, rather than 3.

We have seen that by taking advantage of the fact that the measurements were in pairs we were able to show significance where previously this was not possible. It must be emphasised, however, that such differences between rows must have a logical basis, otherwise a large proportion of equal-sized groups could be manipulated to show a significant difference. Thus if one group of weights was, 801, 806, 808, 805, 807 and 803, averaging 805, and a second group was 804, 809, 802, 807, 806 and 808, averaging 806, these could be arranged in array, side by side, thus:

		Difference
801	802	—
803	804	—
805	806	—
806	807	–
807	808	—
808	809	—

From this table it can be seen that there are 6 negative differences which would be significant by the sign test, suggesting that the 2 sets of figures come from different populations. This is, of course, incorrect, for all we have done is to manipulate the figures into pairs that would be unlikely to occur by chance alone. We could just as easily arrange them in the following way: 808—807, 807—806, 806—804, 805—802, 803—808 and 801—809, which gives 4 plus and only 2 minus signs.

Further reading

The two tests given in this chapter cover most of the simple results obtained by controlled experiments that give two sets of measurements and, calculated in a somewhat different way, they constitute the t test and the t test on paired data. In the t test, however, the standard deviation is used instead of the variance so that the final test of probability is given by the square root of the F value when $n_1 = 1$, which gives the values in the table of t.

There are many simple modifications of the analysis of variance designed to meet greater complexities, but these are outside the scope of this book. As an introduction to these complexities I would advise the reader to start with reading the simple explanations given in Moroney's *Facts from Figures*, for not only are his explanations simple, but he dares to use simple numbers in the examples. For applying the analysis of variance to your own work

I would recommend Snedecor's book for he realises the imperfections in experimental design that may be forced on the junior worker through lack of animals.

If it is usual to use the factorial designs on the problem you are working on, then the inexpensive booklet by Yates, on factorial and split-plot experiments, will be most useful. With these more complicated designs it is wise to be cautious of friends who show eagerness to apply the artifice of calculation without bothering to understand the logic of your design.

Summary

In the last chapter we estimated the variance of the error so that the effect of error could be removed. In this chapter we estimate the variance of the error so that it can be used as a yardstick to measure the comparative size of the variance between the columns, or those between rows and columns.

CHAPTER 6

SIMPLE REGRESSION AND CORRELATION

Logical consequences are the scarecrows of fools and the beacons of wise men.

T. H. HUXLEY

We now come to a somewhat different method of logic. If the average erythrocyte count of cows living at somewhere near sea level was 6 million and that of cows living about 9000 feet above sea level was found to be about 9 million, the hypothesis might spring to mind that the erythrocyte count was influenced by the altitude at which the cattle were living. To test this hypothesis samples might be sought at intermediate levels of say, 3000, 5000, and 7000 feet, and it might be shown that even although the cows belonged to different families and were fed on somewhat different diets, there was a tendency for the cows at higher altitudes to have a higher red cell count.

The usual method of inspecting this type of data is to make a dot diagram. To do this the measurement subject to the smaller error forms the horizontal scale, called x, and represents the independent variate. In this example altitude, being a measurement free from biological variation, would be chosen for this variate, whereas the cows' red counts, subject to so much variation other than altitude, would become the dependent variate and would be recorded in an upright scale as values of y, which can be thought of as yield. When convenient these 2 scales meet at the zero for both scales at the left hand corner of the graph.

If the measurement for each cow is plotted as a dot, at the appropriate altitude, the dots may indicate a straight line. If a statistical test shows that this semblance to a line is unlikely to be due to chance alone, then an association between the red count of cow's blood and the altitude at which it lives has been shown by the logical Method of Concomitant Variation. For a conclusion you would have to choose between accepting that a high altitude produced high red counts, or that cows with high red counts are likely to produce a mountain. In this example there can be no doubt which is the cause, or antecedent, and which is the effect, or consequent.

This chapter deals with fitting a straight line to such data by an arithmetic method, and the statistical tests that can be applied to estimate its validity. Where the data are suitable for the application of a straight line they are spoken of as showing a simple regression.

Simple regression

If we fit a straight line using altitude as the independent variate we can use this line to predict the average blood count that we would expect at any given altitude. In contrast, if we wanted to use the red count to predict the altitude we would have to turn the red counts into the x values and the altitudes into the y values; this would probably give a line that crossed the previous one—as we shall see when dealing with correlation.

In trying to fit a line we may make use of the dot diagram and draw a straight line through the dots. If the position of the line is obvious this freehand line is all that is needed, particularly if we believe that the next set of samples will give a line with a somewhat different slope. But where the dots are scattered in an irregular manner one position of the line may appear just as good as another, so that the chosen line will depend on personal bias. Under these circumstances it is necessary to calculate both the centre and the slope of the line.

As a simple illustration of how this is done let us use Fig.100. Here we see that x, the independent variate shows no error for each dot falls exactly over an x value, whereas y, the dependent variate, has had a little error added to it and resembles a cocked-hat curve. To see this you must telescope the items to the y scale and this will show a frequency table of 1, 1, 2, 2, 2, 1, 1; showing a tendency to cluster round the centre.

The first job is to find the centre of the items. This is easy because as we are dealing with a straight line each variate must give an arithmetical progression, and you will remember that in such a progression the average of the first and last terms equals the average for the whole series. Thus for y, $1+7$, $6+2$, $5+3$ and $5+3$, all average 4. And for x, three times $1+3$ all average 2.

Because of this every dot in the diagram is helping to fix the averages of both x and y, so that, whatever the slope of the line, the one thing we can feel fairly certain about is that the line should go through the two averages.

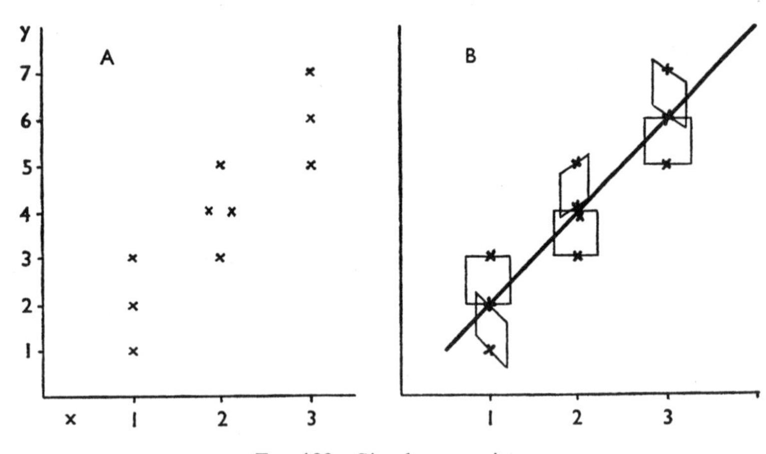

FIG. 100. Simple regression.

METHOD OF LEAST SQUARES

Having fixed the central point we must now try to fix the slope of the line. First let us make a table from Fig. 100 and add a few extra columns that we will require later.

x	y	$(x-\bar{x})$	$(x-\bar{x})^2$	$(y-\bar{y})$	$(x-\bar{x})\times(y-\bar{y})$
1	1	−1	1	−3	+3
1	2	−1	1	−2	+2
1	3	−1	1	−1	+1
2	3	0	0	−1	0
2	4	0	0	0	0
2	4	0	0	0	0
2	5	0	0	1	0
3	5	1	1	1	+1
3	6	1	1	2	+2
3	7	1	1	3	+3
20	40	0	6	0	12

Average for $x=2=\bar{x}$, average for $y=4=\bar{y}$.

From this table we can see that the simplest method of giving a slope to the line would be to put the total for y on top of the total for x, giving $\frac{y}{x}$ or $\frac{40}{20}$, and this would give us a slope in which every unit we added to x would require 2 units added to y.

Although this is the simplest method there is a better one. When dealing with biological distributions one advantage of using the standard deviation to show the range is because it is not the items clustered near the centre that are important but the items at the extremes, and by squaring the deviation from the mean it is possible to give emphasis to these extreme figures. Similarly when trying to find the slope of a line, the items near the middle tell us little, it is the extreme items that set the angle of the line. So that, here too, we use the sum of squares.

From the table you will see that the sum of squares for x is found in the usual way. y is treated somewhat differently in that, having found the deviations from the mean of y, these deviations are multiplied by the deviations of x to give what is called a product. These products are given in the right hand column of the table and it can be seen that the sum of the products amounts to 12.

From the sum of products we work out what is called the coefficient of regression, symbolised as b, and this is done by dividing the sum of the products by the sum of squares for x, giving the formula, $b = \frac{\text{sum of products}}{\text{sum of sq. } x}$, which here is $\frac{12}{6}$, or 2.

In this example it can be seen that we have adopted a more complicated method in order to arrive at the same simple result that we had by the easier method. This is because the example is nicely symmetrical. Such neat designs rarely occur in nature and the more complicated method is used because it gives the line in which the square of the y distances from the line amounts to the least area; in this example each y value either falls on the line or at a distance of one y unit away from it. The squares of these distances are shown in Fig. 100 and it can be seen that the sum of these squares amounts to 6.

You will find that whatever other line you apply will give you a greater total area than these 6 square units. Thus if you put a transparent ruler over the centre point fixed by the averages and move it round clockwise you will see that as the distance to some dots decreases the distance to others increases, for instance instead of being one unit above one dot and one below another dot, you can move the line so that it is only half a division away from one dot. This makes it one and a half away from the complementary dot, so that the sum of the squares of these distances is $\frac{1}{2}\times\frac{1}{2}+1\frac{1}{2}\times1\frac{1}{2}$ or $0{\cdot}25+2{\cdot}25$, giving 2·5 instead of 2. Because this method gives a line that keeps these distances to the smallest sum of squares it is called the Method of Least Squares. If you prefer a mathematical demonstration to an optical one you will find one given by Levy and Preidel, or if you have learned the simplest form of differentiation (from Sawyer or Wesley) then another method is demonstrated by Moroney.

So far we have found that the line must pass through the point given by the average of y and the average of x, or by $\bar{y}$, $\bar{x}$. And the slope is such that every increase of 1 x unit leads to an increase of 2 y units. From this we can draw a line, but to read the values from a graph may be too inaccurate and we therefore require a formula that will enable us to calculate the regression.

The formula used reads; $Y=\bar{y}+b(x-\bar{x})$ which can be translated as follows: the calculated yield (Y) for the chosen value of x is equal to the average yield ($\bar{y}$) plus the regression coefficient (b) times the number of units of x that the chosen value of x is away from the mean of x. It is easier to understand the formula by using it, so that using our example let us assume that we wish to calculate the yields when x is 2·5 and also when it is 1·5.

When: $x = 2{\cdot}5$	$x = 1{\cdot}5$
$Y=\bar{y}+b(x-\bar{x})$	$Y=\bar{y}+b(x-\bar{x})$
$=4+\frac{12}{6}(2{\cdot}5-2)$	$=4+\frac{12}{6}(1{\cdot}5-2)$
$=4+2\left(\frac{1}{2}\right)=5.$	$=4+2\left(-\frac{1}{2}\right)=3.$

Before taking the next step confirm for your own satisfaction that when the coefficient of regression is 3, the mean of x is 1 and of y is 5, that the calculated yield for $x = 7$, is 23.

Significance

If you shake some metal filings from a flour sifter, or a pepper pot, over the centre of some squared paper, the expected result will be that they will cluster towards the centre and thin out at the periphery. Under these circumstances a horizontal or a vertical line would fit as well as any other in trying to depict the distribution of the dots. Quite often, however, the distribution would show a large number of filings in, say, both the top left hand corner and the bottom right hand corner of the sheet so that the line that best fitted the distribution would be sloped and not horizontal. This slope would be entirely dependent on chance and would have nothing to do with a cause and effect relationship.

Where data give a sloped line we therefore want to know if the slope is likely to be due to chance or if it is significant. To test for significance we can again use the analysis of variance, and if we have already found the slope by the method of least squares our job is already half-done.

We must first work out the sum of squares for y. From our table we see that this would be as follows:

$$1^2+2^2+3^2+3^2+4^2+4^2+5^2+5^2+6^2+7^2 = 190$$

$$\text{False s.sq.} - C.T. = 190 - \frac{40\times 40}{10} = 30$$

We see that the total for y is 30. Now accepting the line in Fig. 99 as correct let us find the sum of squares for y based on this line. Firstly, from the regression we know that where $x = 1$, $y = 2$. So that the three values would give $1-2$, $2-2$, and $3-2$. Secondly, where $x = 2$, $y = 4$, so we would have $3-4$, $4-4$, $4-4$, and $5-4$. Similarly, with $x = 3$, so that the complete data would be as follows:

Deviations from the Regression Line

$x = 1, y = 2$		$x = 2, y = 4$		$x = 3, y = 6$	
d	d²	d	d²	d	d²
−1	1	−1	1	−1	1
0	0	0	0	0	0
1	1	0	0	1	1
	–	1	1		–
	2		–		2
			2		

Total sum of squares = 6.

This gives us the calculated total of 6 which we have already seen pictorially in Fig. 100.

We have now found a total s.sq. of 30 for y, and we have found that 6 of these are due to the inherent error of the data, so that the remaining 24 have been absorbed by using a line that was not horizontal. The practical problem is how to calculate the sum of squares for which the regression is responsible when the data is much more complicated than in our example. The solution turns out to be an easy one, for it can be found by simply squaring the top part of the regression coefficient; so that we have

$$\text{sum of squares for regression} = \frac{(\text{sum of products})^2}{\text{sum of sq. for } x}.$$

For our example this is $\dfrac{12 \times 12}{6} = 24.$

In practice, then, all we have to do is to find the regression coefficient and then, by squaring the numerator, find the sum of squares that depend on the slope of the line. By subtracting this amount from the total for y we obtain the sum of squares for error. By this means we avoid the need to measure the position of every dot away from the line.

In the analysis of variance the regression takes one degree of freedom, which you can think of as the single comparison of the sloped line to the horizontal line. The analysis of variance for y will then be as follows:

Source	Sum of squares	D.F.	V.	F.
Total	30	9		
Regression	24	1	24	32
	—	–		
Error	6	8	0·75	

If we look up $p = 0{\cdot}05$ in our F table with $n_1 = 1$ and $n_2 = 8$, we find the figure 5·32. The result is therefore significant and Fisher and Yates' tables show it to be highly significant.

Working example of regression

When using actual experimental data the figures are more difficult to deal with, and this is combated by using short cuts. To see how these work let us examine some tests on red corpuscles as represented in Table 106 and Fig. 105 but keep in mind that the method is exactly the same as that already used; all the difficulty lies in understanding the short cuts introduced to shorten the arithmetic.

If the red cells of the blood are put into weak salt solutions they swell, and if they are put into strong solutions they shrink. Table 106 shows the results for one series of measurements which is depicted in Fig. 105.

The dilutions of salt are in a regular arithmetic progression and therefore supply the obvious independent variate. The average is obviously the half-way point which is 0·76 per cent. NaCl. and

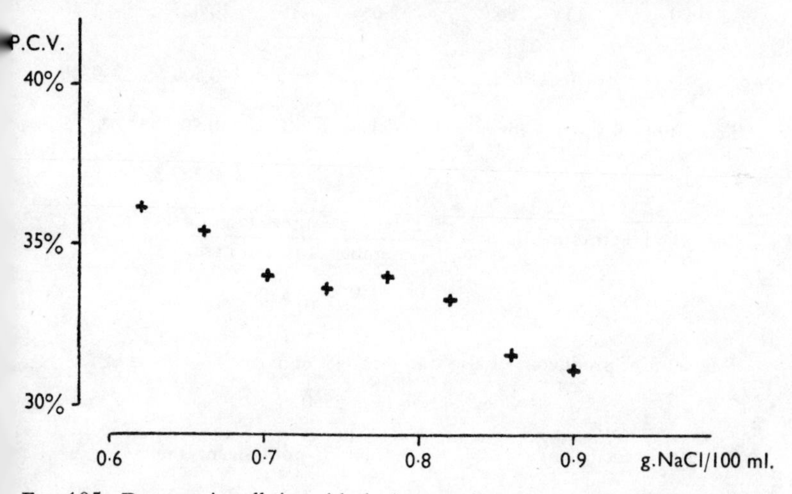

FIG. 105. Decrease in cell size with the increase in concentration of salt solution.

as the series is regular there is no need to find a false sum of squares because it is just as easy to use the true mean as any other. If you wish to get rid of the decimal points then the data can be coded by multiplying by 100.

The volumes given by the red cells centrifuged to the bottom of a tube under standard conditions, is taken as the dependent variate. Here, too, we can get rid of the decimal points by multiplying by 100, but with this data it is worth selecting a false mean, and I have chosen 3400. Because we have taken a false mean we must correct the sum of squares by deducting the usual correction term of $\frac{(T)^2}{n}$ giving $\frac{(-30)^2}{8} = 112{\cdot}5$.

TABLE 106

Variation in the volume of erythrocytes with different strengths of salt solution.

Percentage Concentration of NaCl (x)	$(x-\bar{x})$	$(x-\bar{x})^2$	Packed Cell Volume	Coded y	y^2	$(x-\bar{x})(y-\bar{y})$
0·62	−14	196	36·25	225	50,625	−3150
0·66	−10	100	35·45	145	21,025	−1450
0·70	− 6	36	34·50	50	2,500	− 300
0·74	− 2	4	33·70	− 30	900	+ 60
0·78	2	4	33·95	− 5	25	− 10
0·82	6	36	33·30	− 70	4,900	− 420
0·86	10	100	32·50	−150	22,500	−1500
0·90	14	196	32·05	−195	38,025	−2730
	0	672		− 30	140,500	−9500

$$\text{True sum of squares for } y = 140{,}500 - \frac{(-30)^2}{8} = 140{,}500 - 112{\cdot}5.$$

$$= 140{,}387{\cdot}5.$$

$$\text{Correction Term for products} = \frac{\text{Total for } x, \text{ times total for } y}{\text{number of paired items}}$$

$$= \frac{0 \times -30}{8} = 0.$$

$$\text{True sum of products} = \text{False sum} - \text{correction term}$$

$$= -9500 - 0 = -9500.$$

$$\text{Regression coefficient} = \frac{-9500}{672} = -14{\cdot}14 \text{ (Codings cancel out).}$$

$$\text{True mean for } y = \text{False mean} + \text{error} = 3400 + \frac{(-30)}{8} = 3396{\cdot}25 \text{ coded.}$$

$$\text{Decoded} = \frac{3396{\cdot}25}{100} = \text{apprx. } 33{\cdot}96$$

Regression $Y = \bar{y} + b\,(x - \bar{x})$
$= 33{\cdot}96 + (-14{\cdot}14)\,(x - 0{\cdot}76)$

Analysis of Variance

Source	Sum of squares	D.F.	Variance	F.
Total for y	140,387·5	7		
Regression	134,300·6	1	134,300·6	132·4
Error	6,086·9	6	1,014·5	

To find the products we multiply the true deviations from the mean of x by the deviations from the false mean of y. This introduces a new trick, which is: that when you use a false mean for x and y the sum of the products can be corrected by subtracting a correction term consisting of the total of x times the total of y, divided by the number of measurements. In this example we will therefore have $\frac{0 \times -30}{8}$ which is zero; thus by using the correct mean for the independent variate we have avoided any need to correct the sum of products.

It is no good just reading about short cuts without trying them out for yourself, so that it is useful to confirm that this one will work by using a false mean of 0·60 for the salt concentration. You will then find that all your deviations in the second column are positive and total 128 instead of 0; the column of products will begin 2×225 giving 450, and will total -9980. The correction term will be $\frac{128 \times -30}{8}$ which amounts to -480, so that the true sum of products will be: $-9980-(-480)$ or -9500 as before.

We have now got 672 as the sum of squares for x and -9500 as the sum of products. In coding both x and y were multiplied by 100 so that this cancels out and the regression coefficient will be $b = \frac{-9500}{672}$ or $-14{\cdot}14$. The complete regression formula is shown in Table 106. Note that as the line falls from left to right we have got a negative regression.

When drawing in the line over the dot diagram it is best to cut down the error in drawing by taking points at a good distance from the centre, even if this means extrapolating the line beyond the limit of the dots. As an example, with our data the average is 0·76 and the extreme x values are 0·62 and 0·90, so that easy points to calculate would be 0·56 and 0·96. If these points are correctly calculated then the line is checked by the fact that it passes through the average of both x and y. The calculations would be as follows:

Where $x = 0{\cdot}56$

$$\begin{aligned} Y &= \bar{y} + b\,(x - \bar{x}) \\ &= 33{\cdot}96 - 14{\cdot}14\,(0{\cdot}56 - 0{\cdot}76) \\ &= 33{\cdot}96 - 14{\cdot}14\,(-0{\cdot}2) \\ &= 33{\cdot}96 + 2{\cdot}828 \\ &= 36{\cdot}79 \end{aligned}$$

Where $x = 0{\cdot}96$

$$\begin{aligned} Y &= 33{\cdot}96 - 14{\cdot}14\,(0{\cdot}96 - 0{\cdot}76) \\ &= 33{\cdot}96 - 14{\cdot}14\,(0{\cdot}2) \\ &= 33{\cdot}96 - 2{\cdot}828 \\ &= 31{\cdot}13 \end{aligned}$$

The analysis of variance, as usual, can be carried out without decoding the data. The total sum of squares for y is 140,387·5, that absorbed by the regression is $\frac{(9500)^2}{672}$ or 134,300·6, leaving 6086·9 as the error (or the sum of squares for the distances from the dots to the calculated line). The 8 paired measurements give 7 d.f., 1 of these is taken up in comparing the calculated line to the horizontal one, leaving 6 for the error. F, with n_1 as 1, and n_2 as 6, is 132·4 and Fisher and Yates' tables show that this is highly significant. The assumption that the strength of salt solutions affects the size of erythrocytes has therefore been sustained.

We have calculated that, theoretically, each 1 per cent increase in salt solution decreases the packed cell volume by 14·4 per cent. But a rise of as much as 1 per cent would mean extending the line much farther than the experimental data justifies, so that we must modify this and say that, over a certain limited range used in the experiment, it was found that each increase of 0·1 per cent. in the concentration of the salt solution decreased the packed cell volume by 1·414 per cent.

CORRELATION

The correlation coefficient is a simple and useful index for finding out if there is a significant association between two variates, but, as with the chi-squared test, it can merely show association for it is often not possible to know if either variant is an antecedent or a consequent. Assumptions lead to the misuse of the test. One story illustrating the misuse of this test concerns an enthusiast who worked for a city council. One day, to his horror, he discovered a highly significant positive correlation between the import of boxed dates and the number of illegitimate births in his city. Even more alarming was the fact that they both showed a steady increase over the past few years. Having stopped his own family eating this fruit he was trying to get the sale of boxed dates entirely prohibited when it was pointed out to him that both the sale of fruit and the numbers of illegitimate births varied with the size of the population of the city, which after some fluctuations was now showing a steady increase.

You may be pleased to learn that it is not necessary for you to accept this story as true, but it is worth remembering because it does point out that you are never really sure where you are with correlations. With a regression the independent variate is under your control, but with a correlation you can neither arrange the number of dates to be eaten nor the number of children to be born, and there is always the possibility that they are both dependent on a third and unknown variate. Thus statisticians speak of the

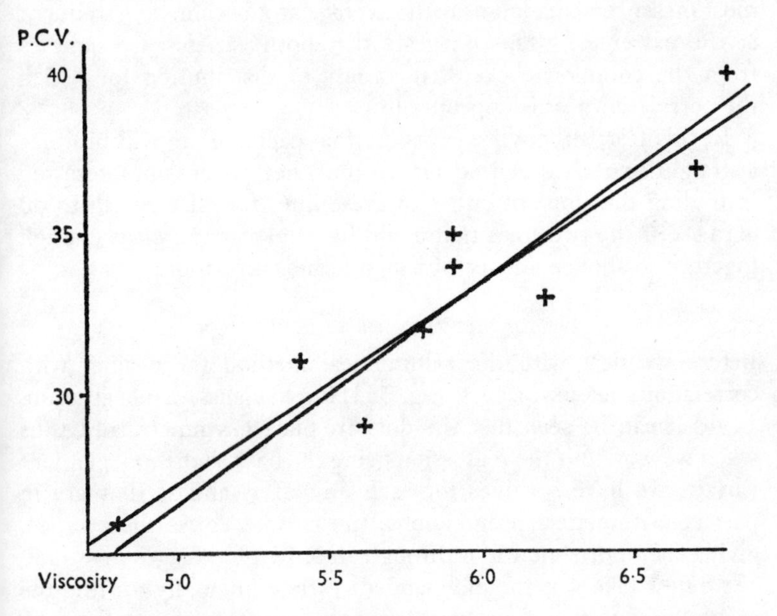

FIG. 109. Showing positive correlation between packed cell volume and viscosity of bovine blood.

highly significant association between cigarette smoking and cancer of the lung, and leave it for others to say that one is the cause of the other.

Having given this warning it can be said that when one wishes to have an objective assessment of the association between two variates, the test is a very useful one and we shall see how it is used.

Suppose we have an hypothesis that depends on the probability that the more red cells there are in a blood sample the more viscous the blood will be. To check this association we spin the blood, as we did in the previous example, in order to get the packed cell volume, and use another part of the same sample to measure the viscosity.

From this experiment we obtain the result given as a dot diagram in Fig. 109. In this instance it does not matter which attribute you take as x for they both have the same type of distribution, thus if you telescope the dots horizontally to the right, as we did before, or vertically to the base, you will see that most measurements fall near the average and become less frequent at the extremes. This suggests that both variates are samples from the common cocked-hat variety of distribution for which the correlation test is appropriate.

In this test, although we assume that each variate will produce a straight line when plotted against the other, we are not interested in finding the slope of either of these lines for all we wish to do is to see if the two lines that could be produced are close enough together to show evidence of a significant correlation.

Demonstration of correlation

Before we deal with the arithmetical method for dealing with correlations let us inspect Fig. 111 for a visual demonstration. In (a) it can be seen that the dots are placed symmetrically, thus when we work out the regression using the base as the independent variate, we have 5 values for each unit of x, and as they are in perfect arithmetical progressions, the middle dot is the average, giving us a horizontal line through the row of averages of $y = 3$. If we now take y as the independent variate (mentally turning the graph on its side if we wish to keep the convention that the independent variate forms the base) we find the same conditions so that for every value of y we have $x = 3$; this gives a straight line at right angles to the first. Under these conditions, when the two regressions are at right angles the correlation coefficient, which is called r, is zero.

Now let us remove the 3 dots in the top left hand corner and the complementary dots in the bottom right hand corner, and see the effect produced. Dealing first with x, we see that for x equals 1 the average is the middle point of the 3 dots, for 2 the average is in the middle with 2 dots either side, and with 3 the line again runs through the middle dot. Thus, by eye we can again draw the regression for x; in the same way, by turning the figure round we can draw in the regression for y. We see that here the lines are no longer at right angles, and that the correlation coefficient has now been calculated as 0·5.

If we continue by wiping out another row of dots from the top left and bottom right corners and again examine the results we shall see that where x is 1, y is 1·5 and at the other end where x is 5, y is 4·5. Turning the graph the other way up the same relationship is shown for y, but except for these small disagreements

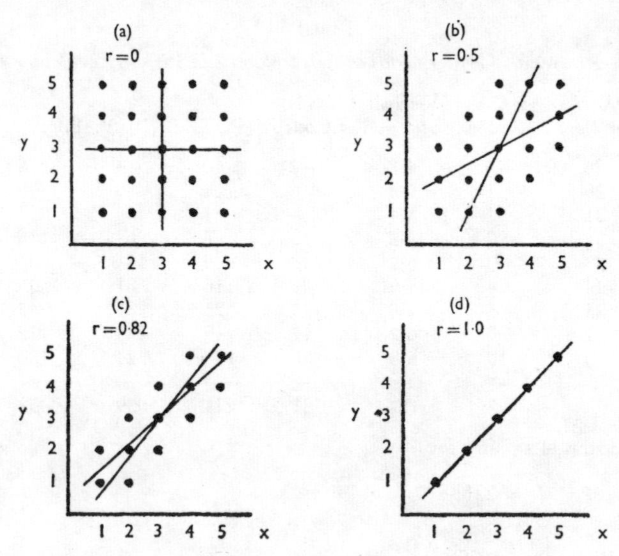

FIG. 111. Strength of correlations.

the averages for the other values are identical. As these lines are taken as straight lines then they must overlap a little and this slight divergence is recorded by the correlation coefficient amounting to 0·82.

Finally, if we remove an additional row of dots from the top left and bottom right positions, we are left with a single row in which the regression for x is exactly the same as the regression for y. In this case there is only one line, the correlation is perfect, and $r = 1$.

The perfect correlation is therefore $r = 1$ and this is 100·0 per cent correlation and no research worker, however enthusiastic, should be able to make it more. If he does, he should check his arithmetic and inspect his approximations to see if perhaps he has approximated too many 0·5s to 1s.

Working example of correlation

We have seen the principle behind correlation, let us now apply the arithmetic to our example of the possible correlation between the amount of red cells and the viscosity of blood. This calculation is recorded in Table 112.

Table 112

Positive correlation between Packed Cell Volume and Viscosity in Bovine Blood.

P.C.V. *Per Cent*	P.C.V. Coded	Viscosity *Aq.* = 1	Vis. Coded	(PCV)^2	(Vis.)^2	P.C.V. × Vis.
26	−4	4·8	−2	16	4	8
29	−1	5·6	6	1	36	−6
31	1	5·4	4	1	16	4
32	2	5·8	8	4	64	16
33	3	6·2	12	9	144	36
34	4	5·9	9	16	81	36
35	5	5·9	9	25	81	45
37	7	6·7	17	49	289	119
40	10	6·8	18	100	324	180
	27		81	221	1039	438

True sum of squares for :

$$\text{P.C.V.} = 221 - \frac{(27)^2}{9} = 140$$

$$\text{Viscosity} = 1039 - \frac{(81)^2}{9} = 310$$

$$\text{Products} = 438 - \frac{27 \times 81}{9} = 195$$

Coefficient of Correlation :

$$r = \frac{195}{\sqrt{140 \times 310}} = 0{\cdot}936$$

Items =9. Degrees of freedom 9−2.

Fisher and Yates' Table VI $p = {\cdot}01$ where $r = 0{\cdot}7977$
$p = {\cdot}001$ $r = 0{\cdot}8982$.

In the first column we have the volume of the red cells after centrifugation and in the next column we have coded these by subtracting 30 from each value. The next column records the viscosities, and the fourth column gives the values after coding by first subtracting 5 and then multiplying by 10 to get rid of the decimal point. Then, dealing with coded values only, we have the squares for each item and the product of the one multiplied by the other: so far the treatment has been the same as with a regression and the columns provide us with the false sums of squares and the false sum of products.

Beneath the columns we work out the true sums of squares by deducting the correction terms, and from the sum of squares, by dividing by the number of items minus 1, we get the variance. Similarly by dividing the sum of products by 8 we get, not the variance, but the covariance.

To calculate the coefficient of correlation we divide this co-variance by the geometric mean of the two variances, which, put as an equation, is this:

$$r = \frac{\text{covariance}}{\sqrt{\text{variance of } x \text{ times variance of } y}}$$

In this example this is;

$$r = \frac{24{\cdot}375}{\sqrt{17{\cdot}5 \times 38{\cdot}75}}$$
$$= \frac{24{\cdot}375}{26{\cdot}040}$$
$$= 0{\cdot}936$$

As both the covariance and the two variances have been divided by 8 this division by 8 cancels out so that in Table 111 we have taken the short cut of dividing the sum of products by the two sums of squares. Note also that although we have coded these figures both averages are in fact quite simple numbers and if you wish to have the practice of working out the correlation it can easily be done without coding.

Significance

Once the coefficient of correlation has been found the significance can be looked up in Fisher and Yates' Table VI. There were 9 dots so that removing one degree of freedom for finding the mean and one for correlation, we have seven degrees of freedom left. From the table we see that where $n = 7$.

$$\begin{aligned} \text{where } r &= 0{\cdot}75 \quad p = 0{\cdot}02 \\ r &= 0{\cdot}80 \quad p = 0{\cdot}01 \\ r &= 0{\cdot}90 \quad p = 0{\cdot}001 \end{aligned}$$

With $r = 0{\cdot}936$ it can be seen that we have obtained a highly significant answer.

In this working example we wished to check our assumption that the larger the number of red cells the greater the viscosity of the blood, and the coefficient of correlation has given us the objective answer that this is a reasonable supposition. In contrast, if our object had been to find out how much increase in viscosity there was for a measured increase in the packed cell volume, then we would have used the regression method with the packed cell volume as the independent variate. Alternatively if we wished to estimate the P.C.V. from the viscosity we would have used the viscosity as the independent variate. These two lines are shown in Fig. 109. Nevertheless, although we are entitled to work out these regressions to find the line, neither of these attributes is really the independent variate for presumably both measurements are dependent on the amount of water in the blood. Therefore, any assumption that one change is the cause of the other is invalid. It is only when we have the independent variate under our control that we can make any deduction of cause.

Dealing with curved lines

Data often gives a curved line and as the method for dealing with simple regression, described in this chapter, concerns straight lines only, this may put the worker into a quandary. The first point to decide is: does this apparent curvature matter? If the line depends for its curvature on just a few items then this may be due to chance, and a straight line method may be the correct one. Alternatively, even if the curve appears to be real is it important? If a worker is only out to justify his hypothesis that, say, an increase in *A* leads to an increase in *B*, then applying a straight line technique to a slight curve may give him the objective confirmation he requires.

Where the decision is made that the curve is important then the most common type of curve in biology is the logarithmic one, and a few selected items that portray the line of the curve can be plotted on semi-logarithm paper. If they give a straight line then a simple regression can be calculated after recording either x or y, whichever is appropriate, as logarithms. If this fails the selected items can be tried on double logarithm paper, and if this is successful both x and y can be turned into logarithms, or an inspection can be made to see if one variate is the square root, or some other root, of the other one. Other papers that can be tried are reciprocal paper and probability paper, both arithmetic and logarithmic.

In the exact sciences the formula of a line may be of the highest importance, but in biology, with its many variables, the exact formula for one sample is less important, for one is seldom sure that the next sample will give the same curve. Where the formula is important then it would be necessary to ask help from a mathematician who was practised at fitting curves.

Regression involving multiple samples

My own work has often involved working out regressions based on multiple samples, and as I had difficulty in finding out how to test these for significance I describe below the method given to me by Dr R. C. Campbell.

We can take as an example an investigation in which the haemoglobin concentration was measured in blood samples taken each week for 5 consecutive weeks from the same 5 kids. And we wish to decide if the apparent increase in concentration is a true upward trend or is likely to be due to chance alone.

The relevant data is recorded in Table 116 and it can be seen that the 5 kids resemble the stage army referred to under the heading of Multiple Samples earlier in the book; thus each week shows the same 5 warriors with some alteration in helmets and weapons.

If we were using the data in Table 116 to show that there was more than a chance difference between the averages for the individual kids using an analysis of covariance, then we would use a total of 25—1 degrees of freedom. But this analysis is outside our province and the only point I wish to draw from it is that in the full analysis we would use 1 d.f. for the regression of the kids as a whole, and 4 d.f. for the regression between the 5 individual kids, for it is these regressions that will be used to test the multiple samples.

TABLE 116

Increase in haemoglobin values (g/100 ml) for 5 kids over period from 4 to 8 weeks of age.

Age		Wk. 4	Wk. 5	Wk. 6	Wk. 7	Wk. 8	Kid Totals	
Coded	x	-2	-1	0	1	2		
	x^2	4	1	0	1	4	xy	$(xy)^2$
Kid	1	6	9	11	11	12	14	196
	2	8	10	11	11	12	9	81
	3	7	8	11	10	12	12	144
	4	9	11	12	12	12	7	49
	5	9	10	14	12	14	12	144
Sub. tot.	y	39	48	59	56	62	(264)	614
	xy	-78	-48	0	56	124	54	

S.sq. for regression for 5 kids as a group

$$\text{Sum of products} = 54 - \frac{0 \times 264}{25} = 54$$

$$\text{S.sq. for regression} = \frac{54^2}{50} = 58{\cdot}32$$

S.sq. for individual regressions of 5 kids

$$\text{S.sq. within kids} = \frac{196}{10} + \frac{81}{10} + \frac{144}{10} \ldots\ldots = \frac{614}{10} = 61{\cdot}4$$

Analysis

Source	*s.sq.*	*d.f.*	*V*	*F*	*t*
Regression within kids	61·40				
Regression of group	58·32	1	58·32	75·74	8·7
Regression between kids	3·08	4	0·77		

$p = 0{\cdot}001$

Because the samples were taken at regular intervals we can begin by coding the five weeks as -2, -1, 0, 1 and 2. This will give an average of 0 for x so that no correction term will be required either for the s.sq. for x or for the sum of products.

Dealing with the 5 kids as a group we have the sum of y for each week recorded for each column and these multiplied by the appropriate x value give the products. These products total 54 and this needs no correction. The squared differences of x are recorded in the 3rd row and they amount to 10, and as each column has a frequency of 5 the total s.sq. is 50, so that the coefficient of regression will be $\frac{54}{50}$ and the s.sq. for the regression will be $\frac{54^2}{50}$, or 58·32.

Turning now to the regressions for each individual kid, the sum of xy for each kid is seen in the last column but one. There is no need for a correction term, and together they check the total of 54. The s.sq. for x for each kid is 10, so that the total s.sq. for the five regressions amounts to $\frac{614}{10}$, or 61·.4

The analysis is shown at the foot of the Table and it can be seen that we use the s.sq. of the group regression and the sum of the individual regressions to find the s.sq. for the difference of regression between kids. We then use this difference of regression between kids for testing the group regression, and it can be seen from the Table that this gives the very large t value of 8·7. From the t Table in the appendix we see that even with only 4 d.f. this gives a probability of only 0·001.

Is there a less complicated method of testing these data? Yes, we could overcome the difficulty of using multiple samples by restricting ourselves to using the average for each week. The 5 averages would be 7·8, 9·6, 11·8, 11·2 and 12·4, and the s.sq. for y would be $571{\cdot}44 - \frac{52{\cdot}8^2}{5} = 13{\cdot}87$. The sum of the products comes to 10·8 and the s.sq. for the regression is 11·66, giving the following analysis:

Source	*S.sq.*	*d.f*	*V*	*F*	*t*
Total y	13·87	4			
Regression	11·66	1	11·66	15·83	3·98
Residue	2·21	3	0·736		

The t Table in the appendix shows that for 3 d.f. the answer gives a probability of 0·05, and if this bare level of significance would suffice then this easier method would be acceptable. Where samples are taken on, say, 10 occasions then the use of the average is often quite adequate to show significance, especially where the line is straight.

In contrast, if we had ignored the fact that these samples were from the same kids, and used 25—1 d.f. as we would for 25 kids, then the t value would have been 6·1 with 23 d.f., giving a probability of only 1 in 10^9. This would give us false confidence in this instance but with other sets of data it could obviously produce false answers of significance.

Concomitant observations

Following on from the calculation of regressions comes making use of concomitant observations, and this will be illustrated in the following example.

A bacteriologist was investigating the bacterial flora in the carcases of young pigs up to the 6th week of life. Weighed pigs, each from a different litter, were killed at different ages and a bacterial examination was made and, in addition, the weight of various organs was recorded.

It was noted that scouring (diarrhoea) had occurred in some of the young pigs and the impression was gained that in these pigs the thymus gland was relatively larger in size. As the thymus gland is believed to play a part in the protection of animals against bacterial infection, this possible association was of sufficient interest to warrant a statistical test of significance.

The age, body-weight and weight of the thymus gland of these pigs are recorded in the first three columns of Table 119.

Looking at the recorded weights of the thymus gland in the scouring pigs it will be seen that they vary from 8·9 to 25·1 g. while for the non-scouring pigs the range was 8·5 to 37·6 g. It seems obvious that using these figures it will be impossible to show a significant difference between the groups for both groups show a wide range and these ranges are largely overlapping. It will therefore depend on the use we can make of the other data recorded at autopsy—the concomitant observations—whether we can show that the average difference in the relative weights of thymus in the two groups is likely to be due to chance, or not.

TABLE 119

Age, body-weight and weight of thymus gland of young pigs killed for bacterial examination. (Based on figures from Dr Heather B. Sharpe.)

Age (days)	Body-weight (g)	Thymus Gland (g)	$\frac{(c)\times 100}{(b)}$	(d) Adjusted for Age*	(e)²
(*a*)	(*b*)	(*c*)	(*d*)	(*e*)	
Non-scouring pigs					
7	2356	10·1	0·43	0·27	729
10	2470	10·4	0·42	0·30	900
14	5307	23·4	0·44	0·37	1369
16	4791	18·2	0·38	0·33	1089
17	4564	17·3	0·38	0·34	1156
19	7512	37·6	0·50	0·49	2401
20	4593	9·6	0·21	0·21	441
22	4309	11·6	0·26	0·28	784
24	5472	10·9	0·20	0·25	625
25	6038	10·3	0·17	0·23	529
26	5698	8·5	0·15	0·22	484
29	5840	16·3	0·28	0·39	1521
29	5981	9·6	0·16	0·27	729
31	4706	11·8	0·25	0·38	1444
34	6464	13·6	0·21	0·38	1444
36	5131	9·2	0·18	0·37	1369
				(508)	
Scouring pigs					
17	4451	23·6	0·53	0·49	2401
20	4095	15·6	0·38	0·38	1444
24	4621	12·0	0·26	0·31	961
25	6860	23·3	0·34	0·40	1600
28	5698	25·1	0·44	0·54	2916
31	4536	15·0	0·33	0·46	2116
35	6350	8·9	0·14	0·32	1024
				(290)	29476

* Index of response = −0·012% per day,
Day 20 = zero adjustment

Now the skill of the statistician is in making the best use of all the available data and to make the fewest assumptions while doing so, whereas the research biologist who is unwilling to make an assumption has chosen the wrong job. In this investigation there are two concomitant observations that might help to make a more direct comparison between scouring and non-scouring pigs, and each will need an assumption if we are to use the easier method that is described below.

The first set of observations is the body-weights of the pigs. It seems reasonable to suppose that a large pig has a correspondingly large thymus gland. It is even possible that a pig twice the weight of another will have about double the weight of thymus gland. This is our first assumption, and it may be somewhat inaccurate but the point is, not whether it is absolutely accurate or not, but whether it is sufficiently accurate to remove a large amount of the bias due to one pig having a larger body-weight than another, and to deal with this we introduce an "index of response" (Cox, 1958).

Index of response

By analogy with some other organs of the body we will therefore make the assumption that the weight of the thymus gland is proportional to the body-weight and on this basis work out an index of response. By using this index, instead of having one value (x) for the body-weight, and one value (y) for the weight of the thymus gland, we shall combine these different values not as products (xy), as in a regression, but into an adjusted y value. Following a pattern laid down by pharmacologists we therefore record the weight of the thymus as a percentage of the body-weight, according to the formula:

$$\text{Index of response} = \frac{\text{Thymus wt.} \times 100}{\text{Body-weight}}$$

and this gives the values recorded in the fourth column of Table 119.

Looking at this column it can be seen that there is still little chance of being able to show a significant difference between the groups because there is still a large overlap in the thymus weights. We must therefore take into consideration a second concomitant observation, that is, the age of the pigs.

Age is important because it is known that the thymus gland quickly decreases in size after birth. We cannot estimate the rate of this regression by analogy with any other organ so that to estimate this rate we must make a dot diagram from our data, and this is shown as Fig. 121.

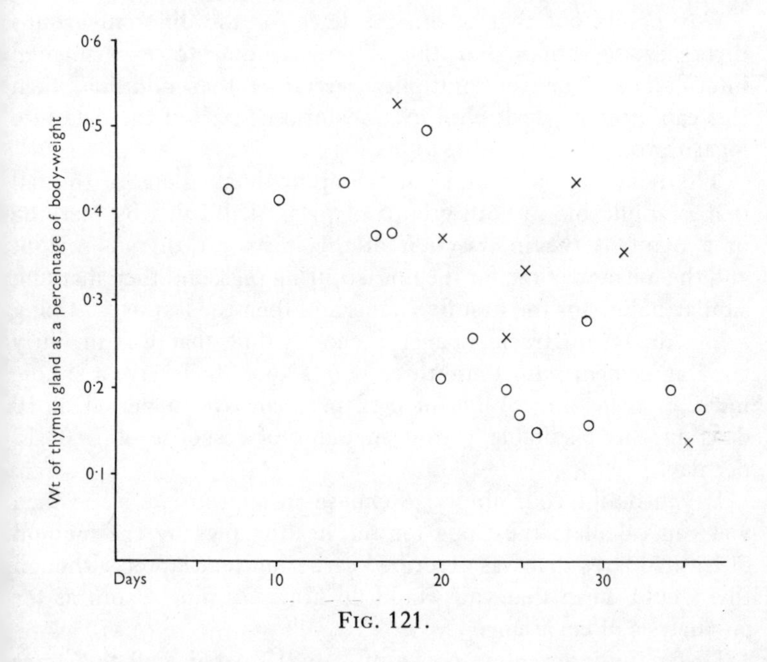

FIG. 121.

The diagram shows the quick fall in the weight of the thymus over the first 36 days of life, and from this diagram we have to make two assumptions before we can supply our method. Firstly, is it possible to assume that these values fall as a straight line?

Using a line drawn on tracing paper it can be seen that a straight line will give quite a good fit. If it did not then we would have to try some of the manipulations already suggested under the heading of curved lines.

Now we come to the second assumption. Is it reasonable to assume that the thymus weights for the scouring pigs fall along a straight line which is parallel to that of the non-scouring pigs? That is, do those of the scouring pigs fall at the same rate but have each had something added to them? It can be seen from the diagram that it would be difficult to say what the correct line for the scouring pigs would be, but that the slope given by the non-scouring pigs would give about as good a fit as many other lines and will not introduce a large amount of error.

Cox points out that where the lines for the different groups diverge, suggesting that the difference due to experimental interference is one of multiplication rather than addition, then this can often be dealt with by transforming part of the data into logarithms.

The next step is to find out the approximate slope of the fall that is applicable to both groups of pigs. I did this by marking in a point at the intersection of the mean for thymus weight and the mean for time for the non-scouring pigs, and then marking similar points for the first five items and then the last five. Using a line drawn on tracing paper I chose a line that was in fairly good agreement with these three points and which gave a simple index of response. A fall of 0·12 per cent over a period of 10 days appeared suitable, giving an index of response of $-0{\cdot}012$ per day.

If you doubt your ability to choose an appropriate slope then you can calculate the slope for the healthy pigs by the method of least squares that was described earlier in the chapter, although this would mean that you would do nearly as much work as for an analysis of covariance.

Using the index of response of $-0{\cdot}012$ per day all the items can be adjusted for a chosen date. For easy calculation I chose day 20. Hence, because the first item was 0·43 per cent at day 7, which is 13 days ahead of day 20, we subtract $0{\cdot}012 \times 13$, giving $0{\cdot}43—0{\cdot}156$, which rounds off to 0·27 per cent. The next item of 0·42 per cent at day 10 gives $0{\cdot}42—0{\cdot}012 \times 10 = 0{\cdot}30$ per cent. Jumping to the final item we have 0·14 per cent for day 35, giving $0{\cdot}14-(-0{\cdot}012)$ (35—20 days) $=0{\cdot}14+0{\cdot}18=0{\cdot}32$. These adjusted values are recorded in the fifth column of the Table.

Having used the first index of response to adjust for the different weights in different litters and the second index to adjust for the difference in ages we can now carry out a test for significance on the two groups. To do this we can multiply the adjusted items by 100 to get rid of the decimal points, and then square each item, and these squares are shown as the 6th column in the Table. Using these adjusted items, we have the following analysis of variance.

ANALYSIS OF VARIANCE ON ADJUSTED ITEMS

From Table 119

Total s.sq. $= \text{items}^2 - \text{main C.T.}$

$$= 29476 - \frac{798^2}{23} = 29{,}476 - 27{,}687$$

$$= 1{,}789$$

S.sq. between groups

$= \text{C.T. non-scouring gp.} + \text{C.T. scouring gp.} - \text{M.C.T.}$

$$= \left(\frac{508^2}{16} + \frac{290^2}{7}\right) - 27{,}687$$

$$= 456$$

Source	*S.sq.*	*d.f.*	*V*	*F*	*t*
Total	1,789	22			
Between treatments	456	1	456	7·18	2·68
Residue	1,333	21	63·5		

$p = 0{\cdot}02$

Hence by using concomitant observations we have shown that the relative difference in size of the thymus gland in the scouring pigs is unlikely to be due to chance alone.

Note that we did not put much effort into trying to find the most accurate line; all we were trying to do was to choose a slope that would remove most of the bias introduced by the fact that the 6 youngest pigs were free from diarrhoea. On the other hand, if you are doubtful of your ability to choose a suitable slope, or if you wish to build up information on the rate of regression of the thymus gland, then you could calculate the regression by the method of least squares that was described earlier in this chapter. If you did that using the items from both groups you would get a regression coefficient of $-0{\cdot}0103$, which, contracted to $-0{\cdot}01$, would give you an easier index than the one I chose.

ANALYSIS OF COVARIANCE

Not only is it possible to calculate the slope by the method of least squares but, as we saw when we tested significance of a simple regression, we can calculate the variance of the items as if they were surrounding a horizontal line instead of a slope by subtracting the sum of squares obtained from the formula $\frac{(\text{covariance})^2}{\text{s.sq. for } x}$ from the total s.sq. for y.

This method also contains additional refinements and does not require us to make assumptions, nor to adjust every item individually. This method, called the analysis of covariance is to be found in all books on applied statistics. If, however, you rarely require such a test then you will be far more confident that you have a realistic result if you choose and employ an index of response, even although the method is less accurate, but for those practised in it, the analysis of covariance is the more efficient method of dealing with these problems.

Further reading

The next step after simple regression is multiple regression. An example of a multiple regression based on the pig experiment would be a single equation in which the calculated weight of a thymus gland would be predicted from a formula in which one regression coefficient took into account the body-weight of the pig and another took into account the age of the pig.

Methods involving the analysis of covariance include those for showing the difference between two or more groups by one of the following criteria: *a.* showing a difference between their variances; *b.* showing a difference between their means; and *c.* showing a difference between the slope of the lines. These methods are explained in books on applied statistics, such as Mather or Snedecor.

Much useful general information is given on the index of response and on covariance in a book by Cox (1958) which is strongly recommended to those who are familiar with routine statistical methods. Some familiarity is necessary because no examples are worked out completely and it is left to the reader to decide whether he has interpreted the author correctly.

Summary

We have seen how, in addition to adding or subtracting sums of squares for error or treatments, we can also subtract those due to the data forming a sloped line instead of a straight one. This gives us the power of using another method of logic, called the Method of Concomitant Variation, by which we can infer a causal relationship if we find that the experimental increase, or decrease, of one factor, the antecedent, is followed by a proportional rise or fall in another factor, the consequent.

A simpler calculation for correlation does not demand that we have either factor under our control, but is limited to showing an association between the two variates, without indicating that the change in one is the cause of the change in the other.

Where it is known that the weight, yield, age or other measurement of an animal, will modify the effect of treatment to a proportional degree, this influence can be allowed for by using the analysis of covariance, although weaker brethren can usually obtain a satisfactory, but less precise, result by the use of an "Index of Response".

Reference

Cox, D. R. (1958). "Planning of Experiments." John Wiley and Sons, London.

Chapter 7

GOODNESS OF FIT AND NON-PARAMETRIC TESTS

To do a great right, do a little wrong.

W. Shakespeare

In Chapter 1 it was stated that the exact test was a non-parametric test, but no explanation of this term was given. In this chapter an attempt is made to explain the terms parametric and non-parametric, and to introduce other non-parametric tests.

It has already been stated that parameters are measurements of the population by which we can identify the shape of a population. Thus with a binomial distribution, in which we deal with the frequency of items showing different qualities, our parameters could be the number of items and the chances for and against their occurrence. Once we know these things we can draw the shape of the histogram by using the formula $(p+q)^k$ according to the binomial theorem (see for example, Wesley or Levy and Preidel); where items are very numerous a more complicated formula, similar to that used for the normal distribution, is employed. With the normal distribution the frequency of the different measurements can be calculated from the parameters of the mean and the standard deviation, either by using a formula that includes these parameters, or by using either a table for the normal curve or the table of t, in which the work is done for us. In most tests for significance these parameters are not known to us and we estimate the parameters of the population as best we can from the average and standard deviation of our sample, calling these measurements statistics to distinguish them from the real thing. Samples that are likely to have come from a binomial or a normal distribution are described as parametric.

In this country, before the use of motor tractors, ploughs were pulled by plough horses, and a sample of beasts of burden would have given a cocked-hat curve that would have suggested a parametric distribution. In contrast, in poorer countries a sample of beasts of burden yoked to a plough could include a camel yoked with a bullock, or even the peasant's wife yoked with a donkey. Under these circumstances a sample could include camels, bullocks, horses, mules, donkeys and Mrs. Peasant. This would give such an irregular sample that no practical mathematical formula could be used to identify its shape. This then is an extreme example of a non-parametric population.

To test a sample to see if it is likely to have come from a parametric distribution we test for Goodness of Fit. With a binomial distribution this is easily done by using the χ^2 test. If the result judges the distribution to be binomial the data is spoken of as homogeneous; if the sample is judged as unlikely to have come from a binomial distribution it is spoken of as heterogeneous.

To test a sample to see if it comes from a normal distribution is more difficult, and requires a large sample. But, owing to the efficiency of the analysis of variance—which depends to a large extent on the fact that averages tend to give a normal distribution even where the individual items give an irregular distribution—it is only necessary that the items are clustered together. In order to get this cluster it is sometimes necessary to turn the value of the items into logarithms or square roots, and this will be dealt with in a later chapter.

TEST OF HOMOGENEITY ON OBSERVED DATA

We saw in Chapter 2 that with a large sample the χ^2 test gave a similar probability to the exact test, and could therefore be used as an easy substitute. Where, however, the χ^2 test is used with more than one degree of freedom it tests the probability of the sample coming from a binomial distribution. The following example may help to make this clear.

In addition to red cells, the blood of mammals contains white cells, or leucocytes. These leucocytes can be counted in the same way as red cells by using a haemocytometer. Pattison and McDiarmid (1943) bled anaesthetized guinea pigs from the heart, and formed the opinion that when the needle entered the right ventricle of the heart the number of leucocytes in the blood sample withdrawn was higher than when the sample was obtained from the left ventricle. An abridged table of their results is given as Table 129 It will be seen that the actual number of leucocytes counted under the haemocytometer is recorded in this table although, for example, the top left hand figure of 90 would normally be recorded as 4500 leucocytes per c.mm. of blood. This is because, as we have already seen, the error inherent in a count is halved by counting four times the number of cells, so that it would be very wrong to carry out a test as though we had a sample as large as 4500 when all we had counted was 90.

The results were tested for significance by the late Dr. W. O. Kermack, on the following lines. If the error of making a dilution of the blood is negligible, and if the red cells and white cells are well mixed in the same way as red and white cells in a bag, then the error of sampling will give a binomial distribution. For the first duplicate the results were 90 and 86, and the best estimate we can get as to the true number in the fixed amount of blood is the average of these results, that is 88. Using 88 as the theoretical value we can test the likelihood of the two results by the χ^2 test, thus:

$$\frac{(A-T)^2}{T} = \chi^2 \text{ is here } \frac{(90-88)}{88} + \frac{(86-88)}{88} = 0{\cdot}09.$$

In this special case we can save ourselves work by using another formula to avoid two calculations for each duplicate. This formula

is $\frac{(x'-x'')^2}{x'+x''}$ which gives: $\frac{(90-86)^2}{90+86} = \frac{16}{176} = 0{\cdot}09.$

If this formula is worked out for each pair of duplicates on the left hand side of Table 129 it will be found that the sum of the 10 χ^2 values will be 3·48. Each duplicate provides one comparison, or is split by one wedge, so that each provides 1 degree of freedom, giving 10 degrees. If we look up Fisher and Yates' Tables for the row for 10 d.f. we will find the nearest results are 3·059 for $p = 0{\cdot}98$ and 3·940 for $p = 0{\cdot}95$. Our result is so good that a better one would occur somewhat less than 5 times in 100 trials. It is therefore suspicious, and it might suggest some bias in the operators' minds towards getting a good agreement between the duplicates.

This introduces a new concept, that the χ^2 test can not only show that the results are too irregular for a binomial distribution, but can also show that the results are too near a perfect binomial distribution for chance to have played its proper part, and this sometimes suggests manipulation by the experimenter who is trying to get a " good " result.

In this example, however, there is no suggestion at all that the workers manipulated the results, for the explanation is as follows. The distribution of blood cells in a haemocytometer has often been used as an example of a Poisson distribution. The Poisson distribution is an extreme example of a binomial distribution in

TABLE 129

Duplicate Total Leucocyte Counts on Different Samples of Blood. From same side of Heart				Total Leucocyte Counts in Samples from Right and Left Ventricles			
Blood Sample	1st Count (x')	2nd Count (x')	χ^2 $\frac{(x'-x'')}{x'+x''}$	G.P. No.	Left Vent.	Right Vent.	χ^2
1	90	86	0·09	27	192	270	13·17
2	85	86	0·01	12	226	298	9·89
3	92	76	1·52	21	318	386	6·57
4	98	82	1·42	14	90	122	4·83
5	44	45	0·01	25	162	196	3·23
6	60	62	0·03	26	118	140	1·88
7	105	98	0·24	23	204	230	1·56
8	112	117	0·11	22	168	180	0·41
9	111	114	0·04	24	72	74	0·03
10	148	150	0·01	16	176	172	0·05
	945	916	3·48		1726	2068	41·62

which the odds are very uneven; in this example the presence of 88 white cells in proportion to the large amount of diluent, which has taken the place of the red cells. Further investigation has shown (see, for example, Berkson, Magath and Hurn, 1940) that the distribution in a haemocytometer is not strictly a Poisson distribution because, when using some diluents, the blood cells tend to repel each other and thus modify the discrepancies one would expect to find in a binomial distribution. In this example the diluent appears to have obliterated any errors made in diluting and mixing the blood sample, and, in addition, made the distribution within the counting chamber an unusually perfect example of a binomial distribution.

We have now tested the left side of Table 129 in which duplicate samples were taken from one side of the heart, and we have found the results to be unusually homogeneous. Now we turn to the right side of the table and treat each pair of duplicates in exactly the same way. On this side of the table duplicates show much greater discrepancies, with four χ^2 values above the 3·8 values that should occur only once in 20 results. The total χ^2 values amount to 41·62 for 10 d.f. and from the χ^2 we can see that this is well above the value needed for $p = 0{\cdot}001$. The results are therefore very highly significant that the material is heterogeneous.

To sum up, inspection had shown that when two blood samples came from different sides of the heart then in 9 cases out of 10, the count of white cells was higher in the sample from the right side. Was this due to extra cells being liberated into the blood during its journey through the lungs? Or was it likely to be due to chance alone? Where both samples came from the same side of the heart the error was of the size associated with random sampling but where they came from opposite sides of the heart the error was very much greater than one would expect from random sampling. Thus the difference between the two groups was very unlikely to be due to chance alone.

Later in this chapter you will see that a significant result could be obtained, from the right side of the table alone, by using Wilcoxon's test.

Comparing observed data with theoretical values

Besides comparing the homogeneity of duplicate observed values it is possible to compare the goodness of fit between observed frequencies and theoretical ones. Examples of this are often seen in genetical studies.

For a simple example, if N is a dominant gene for blackness in fowls, and n is a recessive one for whiteness, then mating 2 (Nn) birds could produce the following alternatives: (NN), (Nn), (nN) and (nn). As N is dominant the 2 combinations of N and n will produce black animals: The NN is black, but in addition will only produce black progeny; and the (*nn*) combination will produce white progeny. As there is an equal chance of any combination the progeny should work out as 1:2:1, and we can use this theoretical ratio to judge the results. Thus if the observed results were those given below, we could calculate the theoretical frequencies and test for goodness of fit.

	Observed	Theoretical	$\frac{(A-T)^2}{T}$
Black with all black progeny	9	10	0·10
Black with mixed progeny	21	20	0·05
White	10	10	0·00
			$\chi^2 = 0{\cdot}15$

The total for χ^2 is only 0·15 and the result observed could be split into 3 classes by 2 wedges, so that we can look up 0·15 for 2 d.f. in the χ^2 table. We find that $p = 0{\cdot}95/0{\cdot}90$, a very good fit. When comparing an observed ratio against a theoretical ratio, a very high probability, such as $p = 0{\cdot}99$, is purely the indication of a very good fit, and it does not suggest selection or manipulation.

Non-parametric data

In this chapter we have seen how an hypothesis can be justified, in one instance by showing that one batch of data was homogeneous and another heterogeneous, and in a second instance by showing that the data was significantly homogeneous when compared to a theoretical proportion. We now turn to a problem in which the heterogeneity is unconnected with the hypothesis. This is evident in the example given below, but in order to point a warning the data will first be tested by an inappropriate test.

In Table 132 we have the results of an experiment that was designed to test 2 factors: Does the protection given by a particular vaccine wane over a period of 8 months? And do 3 injections of vaccine give a better immunity than 2 injections? The measurement of protection was the ability of mice to withstand a test dose of the live organism, and the result of each test was recorded as the number of mice killed and the number surviving. It can be seen that there were 10 mice in each test. The 2 right hand columns will be referred to later.

Assume that we decide to test the results by the χ^2 test: there will be more than 1 degree of freedom so that we cannot use Yates' modification. As we cannot use Yates' modification for small numbers, the individual cells contain too few items for a reliable result, so that the best thing to do is to contract the table by putting each 2 months together. This action gives the figures presented in the condensed table.

TABLE 132

Monthly tests for duration of immunity in mice, following : (a) a double, and (b) a triple injection of vaccine.

Months after First Injection of Vaccine	Double Injection		Triple Injection		Difference and Sign	
	Lived	Died	Lived	Died		Rank
1	4	6	6	4	2	$5\frac{1}{2}$
2	2	8	4	6	2	$5\frac{1}{2}$
3	3	7	4	6	1	2
4	1	9	5	5	4	8
5	7	3	8	2	1	2
6	5	5	7	3	2	$5\frac{1}{2}$
7	6	4	5	5	−1	−2
8	4	6	6	4	2	$5\frac{1}{2}$

Condensed Table

	Double				Triple			
Months	Lived	(T)	Died	(T)	Lived	(T)	Died	(T)
1 & 2	6	(8)	14	(12)	10	($11\frac{1}{4}$)	10	($8\frac{3}{4}$)
3 & 4	4	(8)	16	(12)	9	($11\frac{1}{4}$)	11	($8\frac{3}{4}$)
5 & 6	12	(8)	8	(12)	15	($11\frac{1}{4}$)	5	($8\frac{3}{4}$)
7 & 8	10	(8)	10	(12)	11	($11\frac{1}{4}$)	9	($8\frac{3}{4}$)
	32		48		45		35	
χ^2		5·00		3·33		1·84		2·37

Total $\chi^2 = 12{\cdot}54$

Because the sub-total for each row is 40, it is easy to work out the theoretical value for each column; thus, as the first column contains 32 out of 160, each row will have the theoretical value of $\frac{32}{160} \times 40 = 8$. The formula $\frac{(A-T)^2}{T}$ for the different rows of the first column is $\frac{1}{8}(2^2+4^2+4^2+2^2)$ or 5·0. And the succeeding columns have sub-totals of 3·33, 1·84, and 2·37, giving a total χ^2 value of 12·54.

The double injection requires 3 d.f. to give the 4 rows, and the triple injection requires another 3, so that we can look up $\chi^2 = 12{\cdot}54$ under 6 d.f. in the χ^2 table. This table gives us $\chi^2 = 12{\cdot}59$ for $p = 0{\cdot}05$; thus the result is on the border line of significance. Inspection shows that the higher values are for the Double Injection so that it might be advisable to test this side alone. This gives us $5{\cdot}00 + 3{\cdot}33 = 8{\cdot}33$ for 3 d.f. The χ^2 table $7{\cdot}82$ for $p = 0{\cdot}05$ and $9{\cdot}84$ for $p = 0{\cdot}02$; our result is between these at about $p = 0{\cdot}04$ so that the result for the Double Injection is definitely significant.

Now where have we got to? Using the items that measure the duration of protection following a double injection of vaccine, we have tested the result by the χ^2 test and found it to be significant. Would you agree that the hypothesis has now been justified?

Lancelot Hogben in *Mathematics for the Million* records that Euler, defending religion at the court of the Tsaritsa, pronounced the grave statement that " $\frac{a+b^n}{n} = x$, therefore God exists! " His opponent, Diderot, was completely overawed by this display of mathematical learning, and retired hurt. Some experimenters work on similar lines, and having shown that something is highly significant, appear to believe that they have won some kind of victory which will quench all argument. The sad fact is that our χ^2 test here is as pointless as Euler's nonsensical statement.

To see if the protection given by the vaccine wanes over a period of months, the commonsense thing to do is to use some squared paper and plot the number of deaths as a vertical measurement with the number of months as a horizontal measurement. Using this method we can see that the results for the double injection are meaningless. In fact it seems obvious that the organisms used for the test dose must have varied in their number, or their virulence, so much that they cannot be looked upon as random samples from the same culture. Perhaps the broth in which they were grown varied, perhaps the culture lost pathogenicity after the third test, or perhaps, after killing nearly all the mice at the third test dose, the experimenter made sure that all the technical errors would be on the side of " too little " rather than " too much." Whatever the reason, the great variation introduced by this test has made the data non-parametric, and no parametric test, such as the χ^2 test with more than 1 degree of freedom, can be used to test results.

We have seen that the great variation at different time periods has obscured any visible evidence on the duration of immunity, so that any statistical test would be pointless. We can still, however, try to compare the double injection with the triple injection by means of a non-parametric test.

Non-parametric tests

Two non-parametric tests have already been mentioned. These are the exact test in Chapter 1, and the sign test in Chapter 5. Can we use these on the vaccine results? The answer is yes, in fact we can use either.

If we total the monthly results we can produce the following 2×2 table.

	Lived	Died
Double injection	32	48
Triple injection	45	35

An exact test on these figures would be valid whether the frequencies in the individual test were distributed in a binomial fashion, or not. The actual test, however, would be tedious, so that, as there is only one degree of freedom, and as we know that with reasonably large frequencies, such as these, the χ^2 test will give a very similar answer, we can use this quicker test. Such a test gives $\chi^2 = 3{\cdot}6$ and $p = 0{\cdot}06$. Thus the result is not significant, although near to the borderline.

As the equivalent of the exact test does not give significance we could try the sign test. In Table 132 on the right hand side there is a column of differences between the double and triple injections, and we can see that in all instances, except one, the number of survivors is greater for the triple injection. Is this proportion of 7:1 significantly different from a null hypothesis that would postulate the commonest chance finding as half positive and half negative?

To test the proportion of 7:1 we can consult the binomial coefficients given, for example, in Barlow's Tables. For the results of 8 trials we see frequencies running 1, 8, 28 . . . and totalling 256. The cumulative frequencies for no negative results, and one negative result, would be $1+8$ at each end, giving the double sided probability of 18 in 256, or about 7 per cent.

Alternatively, if we lacked tables, we could use the χ^2 test, making use of the formula we used earlier in this chapter, but applying Yates' modification as shown by Snedecor (1956). This gives us:

$$\chi^2 = \frac{(x'-x''-1)^2}{x'+x''} = \frac{(7-1-1)^2}{8} = 3{\cdot}125$$

As this is under 3·38 we know that the result is not significant. Alternatively, if we have Yule and Kendall's Table 4B we can see that $p =$ about 0·078. Hence the estimation given by signs alone is a probability of 7 or 8 per cent, and therefore somewhat below the 5 per cent limit set for significance.

We have now tried the two non-parametric tests we already know and have failed to show significance. The position, however, is very similar to that of the test for guinea pig weights that we met in Chapter 5. In this we had to consider, not only the sign of the differences but the sizes of the differences, and to do that with this data we must apply Wilcoxon's test.

WILCOXON'S RANKING TEST ON PAIRED DATA

Like the exact test, Wilcoxon's test (Wilcoxon, 1945, 1947, 1949) is based on permutations, and these are combined with a process called ranking. If we had 5 differences, such as $+1$, -3, -7, $+300$, $+4009$, we would place them in array, as indeed they are, irrespective of signs. $+1$ would rank as 1, -3 as 2, and so on up to 4009 which would rank as 5. In Table 132 the column of differences runs, 2, 2, 1, 4, 1, 2, -1, 2. In this array there are several figures having the same values, and here the average rank is awarded. The 3 results at unity represent ranks 1, 2, and 3, giving the sum of 6 and the rank of 2 for each result. Similarly, the ranks 4, 5, 6, and 7 are all occupied by 2's, so they each receive the average rank of $5\frac{1}{2}$. Finally, 4 being the highest value receives the rank of 8.

To demonstrate the basis of the test let us go back to the sign test. In this test we saw that the chance of all differences having the same sign was 1 in 256. For a single item to have a positive (or negative) sign the chance was 8 in 256, giving the cumulative probability of 1+8 in 256. If, for example, we were working with pennies, this would mean that there was a chance of 1+8 in 256 that 1 coin, or less would turn up heads. Supposing, however, we dealt with this tossing from another point of view, and numbered a line of coins from left to right. Then, if we made a table of results by adding up the ranks of the coins falling as heads, we would expect the coin ranked as 1 to be the only head once in 256 times, so that this with the chance of 1:256 for no heads at all would amount to 2 chances in 256. A total of 2 would include no heads, coin 1 turning up a head, and coin 2 turning up a head, giving a total of 3 in 256. For a total of 3 we would add to this chance the chance of coin 3 turning up a head, and, in addition both coins 1 and 2 turning up heads, giving a cumulative chance of 5 in 256. When we got to the total of 5, this would cover not only the 7 chance in 256 for a total of 4 but, in addition, 3 possibilities, the rank 5, the ranks 1+4, and the ranks 2+3, making 10 chances in 256. Based on these probabilities Wilcoxon built up a table and published the rank totals that indicated probabilities of $p = 0{\cdot}05$, $p = 0{\cdot}02$ and $p = 0{\cdot}01$.

Turning now to our example, it can be seen from Table 132 that the sign with the smaller total is the minus total of −2. We have already worked out that a total of 2 will only occur 3 times in 256, giving a single sided probability of 1·172 per cent, or a double sided probability of 2·344 per cent. Alternatively, if we use Wilcoxon's table (Wilcoxon, 1949) where the probabilities are given to the nearest whole number, then along row 8, for 8 replicates, we see $p = 0{\cdot}05$ is 4, $p = 0{\cdot}02$ is 2 and $p = 0{\cdot}01$ is 0. Our T value is 2, so that in this table it takes a value of 1 in 50. It can be seen that by using this refinement on the sign test we have justified the hypothesis that the triple injection affords more protection than the double one, and that this result is valid even when the distribution of the differences between the two treatments bears no resemblance to a cocked-hat curve.

NON-PARAMETRIC TEST

COMPARING TWO GROUPS OF DIFFERENT SIZES

The previous example was worked on paired data but, of course, it is not necessary for data to be paired in order to be non-parametric. As an example of unpaired data let us take an experiment on fleece weights, and assume that we have randomised a flock of 21 lambs into 2 groups. One group was injected with thyroxin and at shearing time the decision has to be made as to whether the injection has increased the fleece weights in this treated group.

TABLE 137A. *Fleece Weights*, 10 units = 1 Lb.

Injected Lambs :	42, 42, 46, 48, 32, 62, 46, 58, 62, 54, 64.
Control Lambs :	46, 42, 34, 40, 30, 44, 60, 52, 40, 28.

If you make rough histograms of these weights you will see that they are not obviously samples of a normal distribution, in fact, those in the injected lambs have 3 modes, such as they are, 1 at 42, 1 at 46, and the other at 62. For safety, therefore, this data should be treated as non-parametric.

The non-parametric test to be used is White's test (White, 1952). This is an expansion of Wilcoxon's tests which were limited to paired data or equal groups of unpaired data; its basis is again ranking and permutations. To use this test on the Fleece weight data the 2 groups are displayed in Table 137B with the ranks on the left hand side.

TABLE 137B

Fleece Weights in Array

Rank	Injected Lambs	Control Lambs	Score for Controls
1		28	1
2		30	2
3	32		
4		34	4
5		40	5
6		40	6
7	42		
8	42		
9		42	(8)
10		44	10
11	46		
12	46		
13		46	(12)
14	48		
15		52	15
16	54		
17	58		
18		60	18
19	62		
20	62		
21	64		
			81

Inspection of this table shows that the control group would give the smaller score, so that we write the ranks for this group in the right hand column. Note that for the three 42s, ranking 7, 8, and 9, the average figure of 8 has been used: similarly with 11, 12 and 13. The total score comes to 81.

Looking at White's Table for $p = 0{\cdot}05$, given at the back of this book, we see under the 10th column and along the 11th row, the figure of 81. Hence the result is just significant and the greater fleece weight in the injected lambs is unlikely to be due to chance alone.

This is a very useful test and I suggest that you apply it to the data in Table 85. You will see that the result, like that of the analysis of variance, is close to high significance (total score 158 where the limit for $p = 0{\cdot}01$ is 155).

TRENDS

We have already dealt with simple regression and correlation, in which one or both variates are assumed to be samples from a rough cocked hat curve, and we now come to a simple non-parametric method of assessing data for the presence of a trend, which is applicable even where both variates fail to give a cocked-hat curve. The method given is described by Wallis and Roberts in Chapter 18, and they emphasise that " almost any series, if stared at long and hopefully enough, begins to shape up into patterns and cycles." They suggest that a new personality test might take the form of a dot diagram; on the basis that at least one individual would interpret the presence of an upward trend whereas another would be equally certain of a downward one. Where speculation can play such tricks, an objective test is very necessary.

The test is again based on the ranking/permutation principle. Thus if, for a series of successive items, the data ranked 1, 2, 3, 4, 5, then this would represent one arrangement out of a possible factorial 5, or 120. With double sided probability, measuring the distance from the mode, we would take into account the equally extreme occurrence of 5, 4, 3, 2, 1, giving a chace of 1 in 60, and indicating significance.

The calculations become complicated when the order of values produces mixtures of minus and plus differences, and here, as no table has been published, a simple formula must be used.

To demonstrate the test let us imagine that a helminthologist has administered a vermifuge, and is observing its effect by taking serial samples of faeces and counting the number of worm eggs per gramme—it could just as easily be a bacteriologist taking serial samples from a broth culture and counting the number of organisms, or a zoologist counting the concentration of animaliculae from pond water—and assume that the helminthologist's results were as follows:

Serial Sample	Eggs per gramme	Difference
1	1920	
2	2000	−
3	860	+
4	720	+
5	520	+
6	320	+
7	200	+
8	60	+
9	120	−
10	60	+

Looking at these results we see that the second sample was higher than the first, giving a minus difference, the third was lower than the second, giving a positive difference. Positive differences continue until we compare the eighth and ninth samples when a second negative difference occurs.

The helminthologist wishes to know whether these results indicate that the counts are really becoming less, or whether they are likely to occur through chance.

To calculate the probability we use the formula $K = \dfrac{n-2(S+1)}{\sqrt{\dfrac{n+1}{3}}}$

where n is the number of samples, and S is the smaller number of signs, in this instance the minus sign.

For our example the figures are:

$$K = \frac{10-2(2+1)}{\sqrt{\dfrac{n+1}{3}}} = \frac{4}{\sqrt{3 \cdot 666}} = 2 \cdot 09.$$

We are not using Wallis and Roberts' table of K, but as the formula includes a modification for continuity (similar to Yates' modification) we can look up the probability either in the Table of t using infinity as the number of degrees of freedom, or in Fisher and Yates' Table 1 for the normal distribution. It will be seen from this table that $p = 0{\cdot}04$ and that the result is therefore significant.

Corner test (quadrant sum)

For testing an apparent trend seen in a scatter diagram, Olmstead and Tukey (1947) devised an easy geometrical method which can be demonstrated on Fig. 141.

If we assume that the x values represent time then the items are numbered in their chronological order, and the complementary y values show as height from the base line.

To apply the test we find the median x value, which, for 12 items, will fall between 6 and 7. We draw a vertical line at this point. Then, working from the top we find the median y value and put in a horizontal line that separates the top 6 items from the bottom 6.

With a transparent ruler held vertically we can begin at the left and move to the right, counting items 1, 2, 3 and 4. The next item, 5, is on the other side of the horizontal line so this ends the contribution made by this corner, and we record the total of 4 items. Now starting from the top we count 3, 1 and 2 and then find that the next item, 9, is on the other side of the vertical line, so we record the total of 3. Moving the ruler in from the right we count 12, 11 and 10 and find that 9 is on the other side of the horizontal line, so we record 3. Finally we move the ruler horizontally up the page and record 12, 7 and 10 when 5 falls on the other side.

Adding up our recorded totals we have the sum of 13. We then look up Olmstead and Tukey's table for the significance of this total and find that $11 = 0{\cdot}05$, $13 = 0{\cdot}02$ and $15 = 0{\cdot}01$ which gives the demonstration a probability of $0{\cdot}02$.

If there were an odd number of items because item 12 was missing, then item 6 would be the median for both x and y, leaving equal numbers in each direction. If, instead, item 1 was absent, then, underlining the median values to mark them,

7 with a y value of, say, 2 would be the median for x and 8 with a y value of, say, 3 would be the median for y. In this instance we combine $7x\ 2y$ and $3y8x$ into a new item $2y8x$, which reduces the number of items to 10 and allows us to fix boundary lines with an equal number of items each side of them.

If items 1, 2, 3 and 9 all had the same y values then we would find the score for the top left hand corner by using the formula:

$$\frac{\text{no. of items favourable}}{1+\text{no. unfavourable}} \text{ giving } \frac{3}{1+1}=1{\cdot}5.$$

This method is recorded by Wilcoxon (1949) and by Steel and Torrie (1960) who include an extended significance table.

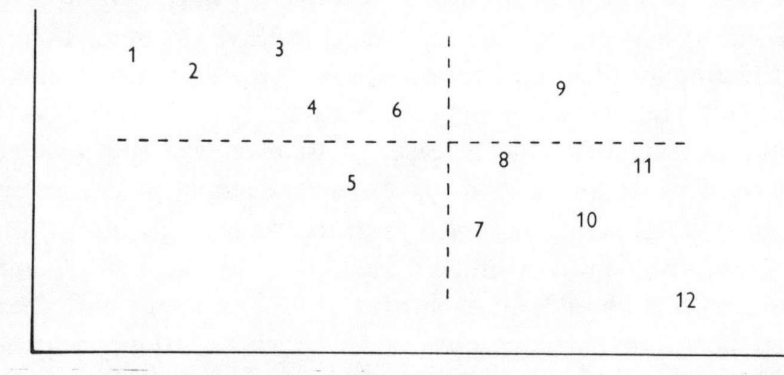

FIG. 141.

USE OF NON-PARAMETRIC TESTS

In this chapter we have seen the use of non-parametric tests. These tests cover practically all the simple questions for which the experimenter requires answers. When the number of items is few, as is often the case in experimental work, they are easy to apply and can be applied even when it is known that the items do fall in a cocked-hat curve. If the result on data distributed as a cocked-hat curve gives definite significance or, on the other hand, indicate that the result could occur by chance once in, say 10 or 12 trials, then no further action is necessary, for the orthodox test on cocked-hat material would not change the verdict of significant, or not significant. If, however, the odds are near to the border of significance then the orthodox test is indicated for, by taking into account the shape of the distribution it can produce a more sensitive measurement, based on the variance, or the standard deviation, which may prove the existence of significance. On the other hand, where the number of items is large, there is no benefit in using a non-parametric test on parametric material, for it would be equally tedious but less sensitive.

SIGNIFICANCE AND FAITH

So far we have spoken of results being significant, or not significant, as though significance had an almost magical property, instead of merely indicating odds of 1 in 20. This was done for the sake of brevity. But when using statistical tests to help our judgement there is no particular virtue in odds of 1 in 20, and our judgement of the result will be influenced by the circumstances, and the faith which they provide. Thus, if we were betting on a coin tossed by a stranger, we might wish to leave the game before losing the amount of money required to reach a significant result. Alternatively, before allowing ourselves to be inoculated with material from an incurable disease, or to receive a special dose of radiation, we might well insist on several experiments giving significant results before submitting to treatment. In both these extreme examples faith would be lacking. On the other hand, although, with 3 animals in each of 2 groups, a result of 1 dead and 2 survivors in 1 group could occur by chance alone twice in 5 times, the knowledge that no previous animal had ever recovered from the test would make the result most exciting, and the treatment given to the group containing the 2 survivors would demand further investigation. Here faith would be strong.

To accept the result of an experiment as being one that justifies the hypothesis is an act of faith by which we form the belief that if we did the same thing again we would produce the same type of result. To say that the result was unlikely to be due to chance alone is only a part of the judgement, for where we are dealing with largely unknown materials, as in biology, the essential factor may remain unrecognised, and it may be present in one experiment and absent in another. Thus the calculation of the odds in a simple experiment is a routine safeguard against over-optimism, but a very highly significant result does not prove that you will always be able to repeat the exact circumstances of the experiment and thus achieve the same kind of result.

This warning seems to be indicated because some workers claim that a result is " highly significant " as though they had won a battle and proved their point beyond contention; others appear to try to force their results to be accepted by claiming that the data has been examined by a "very eminent statistician" and have been found highly significant. As the degree of probability is merely a safeguard to show that the result is unlikely to be due to chance alone, a rough test is often all that the experimenter requires. In this way the difference between him and the professional statistician is similar to the difference between a man who uses his wrist watch to guide him in catching a train, and the horologist who is interested in an accuracy of seconds in a year. If the man, trying to catch a train, gets to the station a little early he is only wasting a little of his time; if he gets there too late, he waits for the next train. Similarly the experimenter who gets an extra high degree of significance has merely been lucky with his material, or has wasted it by using too much; if he gets a non-significant answer, then he must make the next step include a confirmation of the previous one.

Further reading

If the precise probability of a result is not required, non-parametric tests may be used, and where they are simple, they supply surprisingly accurate tools for experimenters who, although they are not prepared to learn much of the theory of statistics, wish to have the safeguard of knowing that their results are unlikely to be due to chance alone. Wallis and Roberts recommend them to all beginners, and Wilcoxon has written a booklet for biologists that consists of these methods alone.

Wilcoxon (1945 and 1947) and Colin White (1952) give full tables of probabilities for their tests; Snedecor, under Short Cuts (Chapter 5) gives abridged tables for both the above tests, and includes Spearman's rank-correlation coefficient method (another ranking test). Wilcoxon's booklet (1949) consists of useful non-parametric tests for two or more groups and the geometrical testing of correlation by the quadrant method. Moroney has a full chapter on Ranking Methods and gives simple formulae for $p = 0{\cdot}05$ and $0{\cdot}01$; Wallis and Roberts include, and advise beginners to use, several non-parametric tests in the chapters on Time Series (Chapter 18) and Short Cuts (Chapter 19), and provide formulae that give the normal curve probabilities.

Summary

These first 7 chapters form an introduction to the type of tools that one finds in the statistical tool chest, and introduces some of the terminology it is necessary to understand before dealing in a more detailed way with experimentation. It can be seen that there are two methods of working out the odds. One method takes no account of the distribution of the data and is called non-parametric: These tests are simple and safe. The other method applies to data giving a rough cocked-hat curve, or data that by some logical transformation can be made to give this distribution; and this method has the advantage of providing a sensitive estimation of probability, even for very complicated designs.

When working out the odds, the statistician must use the most sensitive method and obtain the most precise answer. In contrast, the experimenter may wish to use the simplest method that will show that his results are unlikely to be due to chance alone. But his judgement of the result of the experiment will depend on his faith that he can reproduce these results in the future. As we saw in Chapter 1, this faith should not depend on his desire to defend himself, his friends, or his laboratory, but on his love for objective truth.

Having had a preliminary glance at the different types of observations, and the different methods by which the results can be judged, we can now pass on to a more detailed examination of the anatomy of research method.

References

Berkson, J., Magath, T. B., and Hurn, Margaret (1940). "The Error of Estimate of the Blood Cell Count as Made with the Haemocytometer." *Amer. J. Physiol.*, **128**, 309.

Olmstead, P. S., and Tukey, J. W., (1947). "A Corner Test for Association." *Ann. Math. Statistics*, **18**, 495.

Pattison, I. H., and McDiarmid, A. (1943). "Differences in the Cellular Elements of Blood Obtained from the Right and Left Ventricles of Healthy Guinea-Pigs." *J. Path. and Bact.*, **55**, 217.

Snedecor, G. W. (1956). "Statistical Methods" Fifth Edition, Iowa State College Press, Section 5.8 p. 115.

Steel, R. G. D., and Torrie, J. H. "Principles and Procedure of Statistics." McGraw-Hill Book Co. Inc., N.Y. and London p. 410.

WHITE, COLIN (1952). "The Use of Ranks in a Test of Significance for Comparing Two Treatments." *Biometrics*, **8**, 33.

WILCOXON, FRANK (1945). "Individual Comparisons by Ranking Methods" *Biometric Bull.*, **1**, 80.

—— (1947). "Probability Tables for Individual Comparisons by Ranking Methods." *Biometrics*, **3**, 119.

—— (1949). "Some Rapid Approximate Statistical Procedures." *American Cyanamid Co.*, N.Y.

CHAPTER 8

THE COMPONENTS OF AN INVESTIGATION

> The first essential in chemistry is that thou shouldst perform practical work and conduct experiments, for he who performs not practical work nor makes experiments will never attain the least degree of mastery. But thou, O my Son, do thou experiment so that thou mayst acquire knowledge. Scientists delight not in the abundance of material ; they rejoice only in the excellence of their experimental methods.
>
> AN ARABIAN ALCHEMIST

So far we have been making ourselves familiar with the tools that we will need to cut out our investigation, experiment by experiment, until we are satisfied with its shape. Having seen these tools it is now time to have a look at the investigation as a whole, and see how it is initiated, how one becomes involved in it, and the different parts from which it is built.

The wide perspective of the art of scientific investigation has been recorded by Prof. Beveridge and, from an even wider perspective, the philosophy upon which the investigation rests has been examined by Dr. Agnes Arber. Both these books are of interest to the research biologist but I wish to discuss some of the same points on a lower plane of thought, and to try to act as a more personal guide to this subject.

Prof. Beveridge prefaces his book with a quotation by W. H. George, that " Scientific research is not itself a science; it is still an art or craft." This is obviously true, for research workers are playing with things they do not understand in order to seek the truth about them, but you can play with something you do not understand without seeking the truth, and it is to this form of art that science is opposed. The importance of this distinction is illustrated by Prof. Payling Wright (1950) in the introduction to his book on pathology, where he stresses the fact that clinical medicine, which existed as an art based on the principles of Galen, made very little progress during the hundreds of years these principles were accepted, and it was not until medicine accepted the rational and experimental methods of the abstract physical and biological sciences that the whole tempo of progress in medical research increased rapidly. Even today the choosing of prize bulls for milking herds by means of their conformation, pedigree and the yields of selected progeny, has been gravely questioned by geneticists who appear to be rather rude about the quality of the evidence. Nevertheless, the learning and experience, that produces scepticism in some, produces faith in many others, so that there still remain many erudite professional men and experienced practical men who resent the continually carping mental discipline of scientific method and welcome the more intimate personal interest involved in practising an art, and, to quote T. H. Huxley, baulk at " destroying a beautiful hypothesis with an ugly fact."

Scientific research is sometimes divided into Pure and Applied. The distinction is loose, and it also involves a certain amount of intellectual snobbery. This snobbery is caricatured by the story of the Cambridge don, who on his death bed, thanked God that although he had made several discoveries in mathematics, none of them had been of the slightest practical use.

The difference in method between pure and applied research is negligible and, if research work is looked upon as making a map for other peoples' guidance, then the pure research worker makes an intense study of a small locality to try and find the fundamental principles ordering the lie of the land, while the applied research worker surveys a larger area and is more concerned with suitable positions for roads and water supply. In both types of research merit depends on the continuity of purpose, the imagination shown in overcoming difficulties, and the reliability of the map produced. In contrast, the work is bad if the maps are detailed but inaccurate, or if, through having " grasshopper " minds, the workers have dashed out at the first rumour to examine some locality for the presence of gold, or oil, and returned with an inaccurate sketch map and an uncertain diagnosis. Thus in discussing research work the differentiation between pure and applied is unimportant for either can produce something of the greatest value to the other, but the difference between good and bad research is very important indeed for the value of a map is in its reliability and the sea-serpents depicted in some old-time maps have an artistic role and no direct scientific value.

General pattern of investigation

For the purpose of their books, both Beveridge and Arbor break down an investigation into parts, and I propose to use an amalgam of their classifications, and to discuss an investigation under the following headings.

1. *Choice of an investigation.*
2. *Critical review of the literature.*
3. *Collection of data.*
4. *Analysis of data*
5. *Formulation of hypothesis to explain the data.*
6. *Experiments to test the most likely hypothesis.*
7. *Writing up the investigation.*
8. *Consideration of the possible application of the result to a wider sphere.*

1. *Choice of an Investigation*

How does one acquire an investigation to start with? The conditions are analogous to those of acquiring greatness as given in Shakespeare's *Twelfth Night*, and we can say: Be not afraid of acquiring an investigation: some are born with an investigation, some achieve investigations, and some have investigations thrust upon them.

Under the first circumstance we have the worker who enters a laboratory and is immediately employed in doing the donkey work for an investigation already in progress. As he finds his feet he gradually takes a greater part in the planning and analysis of the experiments; this is the best method of learning to do research work and if the beginner is lucky enough to work for a competent research worker who takes an interest in him, it is the ideal method. Under the second circumstance are those graduates who have obtained some junior position or a scholarship, and hang around hoping each day that they may be blessed with some idea of something inexpensive that they can work on. Finally, in the third circumstance there is the position in which a worker is just told that he will investigate a certain problem; this should only happen to a trained worker, in others it can produce collapse.

Workers may have investigations thrust upon them by employers, patrons, or committees, and as committees are becoming the most important of these sources, a word or two about them would not be out of place. The expense of research work is so great that the cost must be borne more and more by the government. In this country, this means that the money must be obtained from the Treasury, and the Treasury is a body that will not willingly part with its money to anyone but a trained administrator. The trained administrator appears to believe in two things, central control and co-ordination. By central control he raises a pyramid with himself at the top, and by co-ordination he invites many politically important people, with only a vague interest in the subject involved, to sit on committees that supervise investigations.

The true function of the committee was recorded by Topley (1940) and this view is sufficiently important to be repeated. Prof. Topley stated that " Committees are dangerous things that need the most careful watching. I believe that a research committee can do one useful thing, and one only. It can find the workers best fitted, and most anxious, to attack a particular problem, bring them together, give them the facilities they need, and leave them to get on with the work. It can review progress from time to time, and make adjustments; but if it tries to do more, it will only do harm."

Prof. Topley tried to augment this limited power of committees by introducing conferences of the actual workers on a particular subject, but without his presence these conferences were changed back into committees. His idea that committees should know which workers were suitable for certain jobs was a little optimistic and there are still committees who choose workers who delight in uncontrolled experiments, or who choose workers devoid of any knowledge or interest in sampling methods to carry out surveys, with the result that they are later presented with a most interesting set of personal reminiscences, unspoilt by arithmetical facts.

The selection of men for research work remains an art, and the candidate is often subjected to such searching questions as to where he went to school, what games he played and why he wishes to become a research worker. As Topley has pointed out, industry cannot afford the inefficiency of science, so that having made use of statistical tests for several years, industry is now interested in scientific methods of choosing the best worker for the job. Until such a study is made on the attributes of a research worker one can perhaps bear in mind the comments of a certain Japanese general. As I remember it, this general when asked what sort of person he liked on his staff, replied that, where possible, he liked to get intelligent and energetic officers but, as these were hard to get, he was willing to take intelligent men who were not energetic. If they were forced on him, he could find some sort of use for men who were neither intelligent nor energetic, but Heaven forbid that he should ever have anyone who was both unintelligent and energetic! In research work it is these people who buzz round in excited circles creating two problems where previously there was only one.

A good interpretation of the qualities required by a research worker was given by Francis Bacon (1561-1626) when he recorded his own attributes in the following words:

> I found that I was fitted for nothing so well as for the study of truth; as having a mind nimble and versatile enough to catch the resemblance of things (which is the chief point), and at the same time steady enough to fix and distinguish their subtler differences; as being gifted by nature with desire to seek, patience to doubt, fondness to meditate, slowness to assert, readiness to consider, carefulness to dispose and set in order; and as being a man that neither affects what is new nor admits what is old, and hates every kind of imposture. So I thought my nature had a kind of familiarity and relation with Truth.

Yet, perhaps, even more important than the selection of a new worker is his early training, for once a worker has acquired a sloppy technique and bases both his hypotheses and his conclusions on impressions and observations that lack either a direct control or sufficient numbers, his condition is almost incurable, and he may fiddle on for the rest of his life without obtaining one decent clear-cut result.

2. *Critical Review of the Literature*

The most important word here is the use of the word critical; if you are going to accept everything every author says without examining his experimental data, then Nature will appear a very wondrous thing. Alternatively if the author gives no experimental data he will have a good reason for this, and the article will not be worth much study. Some people appear to feel that the farther back in history you go, the better, and complain that things are discovered and re-discovered just because workers neglect reading the older articles. Often, however, it will be found that the older records are lacking in value because, for example, the organism they were dealing with had not then been classified into types, or the materials they were using were not the pure manufactured modern material but a substance contaminated with many impurities.

In other subjects the literature is so vast that it would be quite impossible to read all of it. Even here, although reading other people's mediocre experiments is a dull business compared to planning one's own brilliant researches, and admitting that reading can be carried to excess, you must read enough to understand the subject you are dealing with. The temptation to do the experiment first and read up the subject just before you write up your own investigation, is a real one, but it is very awkward when you then discover an article which tells you that you have just used a most unreliable method.

For many articles a glance at the summary will be sufficient but where the article is of direct interest to the investigation it must be scanned. A good way of doing this is to simulate the experiment on paper, perhaps drawing a circle for each animal, and indicating their treatment and their fate with ticks and crosses, or by arrows pointing in different directions. By this method you really understand what has been done, so much so that you may even feel that the writer has drawn a wrong conclusion, and you can form some opinion of the reliability of the evidence.

Card Index System. With most people there is little value in reading the literature unless some permanent record is made of the important points. The failure to make notes can result in hours being wasted in looking up the wrong journal for the correct date, or the right journal for the wrong date. One method of recording essential points is to use a card index system. My own preference is for the small $5'' \times 3''$ cards, because they can be carried in an envelope in the pocket, and are therefore always available. In addition to the alphabetical markers one can buy additional plain markers, and these can be used for subheading particular subjects. Thus, if you were interested in the working of the heart, then behind the *H* index you might use another marker labelled Heart, and the first card behind this would be numbered, not *H*1, but *Hrt.* 1. Many people wish to remember useful techniques, so that in addition to the *T* marker, there could be a marker labelled *Tech.* Thus, if the articles on the heart gave a useful technique for counting heart beats, then on the *Tech.* marker you could put " Counting Beats, *Hrt.*1 and thus cross-index your first entry.

Remember that, if you are going to quote from the card index, you must not only record the journal, year, volume, and page, but also the author's name and initials; missing or illegible initials may necessitate looking up the article again. If you are concentrating on one particular subject, then it might be advisable to use punched cards; no elaborate machinery is necessary, and the cut cards remain behind as the others are withdrawn with, say, a knitting needle.

3. *Collection of Data*

This is a most important part of a new investigation, for it brings you into touch with the material that is to be investigated and permits you to get the right perspective on the problem. This collection of data is rather similar to the dull detective routine of the " Where were you on the night of the 14th? " type. It is also similar to searching a house for a secret room; if you measure the height, length and breadth of each room so that you can compare it later with the external size of the house, and work systematically through the house, then you must find that there are certain places in which the secret room could exist. In contrast, if you dash enthusiastically round the house, peering into a room here and another one there, trying to estimate its size and to imagine the position of a secret room, then you will have a great deal more fun but after a few days you may have pulled down quite a lot of the house, and then have to consider doing the job systematically with even less enthusiasm for this method than when you started.

Another point is that if we measure the length and the breadth but don't measure the height because we don't want the bother of dragging the steps round with us, or measure any two dimensions without the third, then we may completely fail to show the essential discrepancy. Further, with an actual scientific investigation, we are not always sure which measurement is which, so that all we can do is to try and measure everything that we know how to measure in the hope that three of the measurements will represent height, length and breadth.

This method sounds tedious and irksome but it is carrying out Hughlings Jackson's dictum on pathology, that "the study of the things caused must precede the study of the causes of things" and it is surprisingly effective, so that young graduates who have chosen, or been given a subject for investigation, can rely on the fact that properly kept data on almost any biological material will provide them with some correlation or discrepancy that can provide an hypothesis that will lead to a chain of experiments. Chance can play a big part in an investigation, and it is therefore as well to remember that, although you are looking for a secret room, you may, with a systematic approach, just as easily discover treasure hidden within the fireplace.

4. *Analysis of Data*

Once the measurement of the rooms in the house is complete, then the lengths of the various rooms can be added together to see if they equal the length of the building; the widths of the individual rooms can be compared with the total width, and the heights with the total height. Where all these figures are too intricate to grasp then a plan of each room can be cut out to scale and fitted together within the plan of the building. Similarly, in experimental work, by using the analysis of variance, the sums of squares for the columns, the rows and the treatments can be compared with the total sum of squares; or to get a better perspective, these figures can be simplified by depicting them as histograms or as dot diagrams.

5. *Formulation of an hypothesis to explain the data.*

Once the analysis of data has shown an unexpected agreement, or discrepancy, between the measurements, a possible explanation can be put forward as an hypothesis. This is probably the most enjoyable part of research work, and the imagination can be given free rein. Such a wealth of imagination is, in fact, provided that Beveridge records the saying that " No one believes an hypothesis except its originator but everyone believes an experiment except the experimenter." In general the most highly imaginative hypotheses seem to be associated with the most unreliable data and the most imaginative I have seen was based on a discrepancy in the data for which a more simple explanation would have been the assumption that one instrument was incorrectly standardised.

Long experience has shown that the best hypothesis to choose is the most simple one that will explain the facts. This aphorism is known as Occam's razor. It was formulated by William of Occam (Ockham, in Surrey) as long ago as the fourteenth century, and even at that time the idea was not new. The history of science abounds with really important discoveries based on sound data and simple hypotheses and Beveridge gives an entertaining account of the attributes of a working hypothesis based on that of Christopher Columbus, which brings the value of the hypothesis to its correct proportion, viz.: that it serves as a basis for a good experiment.

A reasonable hypothesis may not only lack novelty but may have been tested before. If you feel that you have a sensible hypothesis and you are told that " it has all been done before " there is no reason why you should not try it again, provided that you are prepared to do it more carefully than it was done by previous workers. With better instruments, less contaminated materials, or better selected organisms, there is no reason why you should not only be successful where others have failed, but also discover the reason for their failure.

6. *Experiments to test the most likely hypothesis.*

The experiment is the principal weapon of the research worker and it is here that his imagination and ingenuity, together with his knowledge and experience, have full scope. The difference between the scientist and the " crank " centres on the experiment, for the scientist having formed an hypothesis has the imagination to find experiments that will prove it right or wrong, while the crank will either choose an hypothesis that nobody can ever prove right or wrong, or will carry on ineffective observations which will give some support to his first hypothesis providing a second, or even a third hypothesis is accepted. Thus the true scientist may show imagination in his hypothesis but certainly shows it in his experiments, whereas the crank shows it only in his hypothesis.

An experiment to test the hypothesis should not be a mad scramble to " skim the cream off the top," undertaken without a sound knowledge of the materials or organisms concerned, and followed by a rush into print. A good investigation is a chain of experiments proceeding step by step over a period of years. The time period for higher degrees suggests that the minimum time for a simple investigation is about three years but, in fact, at the end of this time the worker usually knows just about enough of his subject to begin a useful investigation on it.

The idea that by jumping from subject to subject, trying out the latest fashionable material or method without preliminary investigations, will end in a lucky break that will bring fame, is not supported by the history of science. On the other hand it must be admitted that such conduct is often a useful political gambit for it permits a person to pretend to an expert knowledge of many subjects which may lead him to a better financial position; fortunately it often also removes him right out of the laboratory.

Speaking generally all the easy problems in science have been solved and those that remain do so because their solution requires work, expert knowledge and perhaps a bit of good luck. Before tackling them, therefore, you need to have the best tools that you can get and to learn how to use them. Thus, before starting out on the crucial experiment which will, or will not sustain the hypothesis, or in other instances even before thinking of an hypothesis, the worker will undertake some preliminary investigation to become familiar with the material he is going to use, its effects on living tissues, and the accuracy with which he can measure these effects.

Testing Materials amd Methods. The first thing you want to be sure of is that you are using the correct material and that you have a reliable test by which you can recognise it when you have got it, and also measure the amount present. You may also want to know the conditions under which it is most active and the substances and conditions that are antagonistic to its action.

Similarly, if you are going to work with a certain type of organism you do not just start work with the odd strain that you have in the cupboard but you start to collect strains from other bacteriologists, and start choosing the strain that is best for your purpose. To do this you may try out the recognised tests for it and discover which medium it will grow on best and when the toxin or the bacterial growth is at its height. You can then go on to see which strain it is easiest to type serologically. With a new substance or a new strain of organism, even this preliminary will be measured in months or years rather than weeks.

The advice to become familiar with the material may seem so obvious that it is futile, but be warned that it is dreadfully easy to believe what you read printed on the label of a bottle, or typed on the label of a culture. If by chance you do use the only batch of material that is inert or contaminated, or the only strain of organism that has mutated, remember it is not your bad luck but just your bad technique. Further, remember that it is quite possible to do three or four experiments and find them leading you in a novel direction before you discover you are using this faulty material. In feeding experiments remember that the value of crops vary. Hay or oats from one place in one year may be very different to that grown in another year or another place. When you say that some artificial food is better than hay, or oats, you must know what you mean by hay or oats. Further, you must remember that what seems good hay to you may not appeal to the animal, thus our goats prefer a dirty brown hay that nobody would choose deliberately.

Tests on Susceptible Animals. Assuming that the practical importance of the work will depend on its importance to man or to the farm animals, then a preliminary step will be to try to test out the material and the technique to be used with it in some inexpensive way so that one can become familiar with the weaknesses in technique that are sure to appear. This often means trying to find a susceptible laboratory animal, or alternatively in subjects like helminthology, trying to work with a similar type of worm that naturally infests a laboratory animal.

Finding a suitable animal to work with is an important part of an investigation if the experimentation is not to become ruinously expensive, and once a suitable species is found the whole technique can be put into practice, including such techniques as the recovery of the material, or its products, from the blood stream or the excreta, and the measurement of the physiological changes it produces, such as its effect on growth rate or haemoglobin production. Finally comes a painstaking post-mortem examination and an examination of any histological changes produced, and these must be based on a comparison with direct controls.

The warning constantly being given is that you cannot argue from one species of animal to another. It is right that we should be constantly reminded of this but that should never stop us carrying out preliminary work on small animals, and if in the end it is shown that they do react in a different physiological manner to man or to other animals, this fact may have even more importance than the object for which the experiment was designed.

There is an ethical side to the use of animals for experimentation, but in a world that catches fur bearing animals in traps with metal teeth in order to supply an additional form of clothing, that, in order to add to the variety of its diet, breeds animals and keeps them in captivity until required for slaughter and castrates them if it will make the meat more tender, and that uses the most painful poisons to kill those rats that have not been invited to share our food, it seems natural that man should use animals for test purposes rather than use other men, women or children, especially where the object is to find the cause and the cure of animal diseases.

This principle seems logical to the man in the street, thus propaganda against the use of animals for experiments cannot bear weight if it consists of simple truth and so it is driven to half truths and quarter truths. Of more value to the ethics of this country than those that oppose the use of animals for experimentation is the work of the Universities Federation for Animal Welfare, which seeks to find and encourage all methods which decrease suffering in animals including experimental animals and wild ones; and *Soldiers and Laboratory Animals*, by Major C. W. Hume should be read by all who have not seriously considered the ethics of using animals for experimentation.

7. *Writing up the Investigation*

Before work is written up it is necessary to at least appear to know why it was done at all, why it was done in that particular way, and the true meaning of the result that can be drawn from the data. It is thus often the most irksome part of the investigation, but as it is also the most important part it is treated at some length in Chapter 14.

8. *Consideration of the possible application of the results to a wider sphere.*

The importance of this application is best shown by an example. Prof. Fleming noticed that a fungus, that had contaminated a plate of media, prevented certain bacteria from growing. He recorded this fact, and commented that this power to inhibit bacterial growth would be very useful if it could be applied to control some human diseases. He tried to isolate the active principle but, having failed to do so, thereafter limited himself to using the product of the fungus to control the growth of susceptible bacteria, and thereby encourage the growth of unsusceptible types which were otherwise difficult to isolate. Under these circumstances there was no glory for the professor.

Several years later Chain working with Florey, discovered how to extract the active principle in a stable form, and it became possible to use this substance, called penicillin, to control human diseases. It was only after penicillin had been proved effective for that purpose that Prof. Fleming was showered with honours. If he had not foreseen the wider scope of this inhibitory substance, and drawn attention to it, penicillin might have remained undiscovered for many years to come.

Further reading

See the books by Beveridge and Arber, referred to at the beginning of this chapter.

References

Hume, C. W. (1958). " Soldiers and Laboratory Animals." *Lancet*, **1**, 424.

Topley, W. W. C. (1940). " Authority, Observation and Experiment in Medicine." *Linacre Lecture*, Cambridge Univ. Press, page 31.

Wright, G. Payling (1950). " Introduction to Pathology." Longmans Green and Co. London, page 2.

CHAPTER 9

THE EXPERIMENT

> Take care of experimental design and the tests of significance will take care of themselves.
>
> D. D. REID

The object of an experiment is to test the hypothesis; the amount of care with which this is done varies with the investigator's state of mind. Every year three or four people are killed by their friends with a bullet from an unloaded revolver. This is because the state of mind in which one examines a revolver before pointing it at a friend, is often very different from the careful examination one gives it before putting the barrel into one's own mouth and then pulling the trigger. This latter type of investigation is convincing and few people will object to having a revolver pointed at them if they have seen you form the hypothesis that it is empty, and then point it at your own head and pull the trigger so that the hammer falls in turn on every chamber. This, then, is a well designed experiment, because when you have carried it out, not only you, but other people, would feel reasonably certain that the hypothesis was correct, and that the gun was unloaded. Alternatively your sorrowing friends would be equally convinced if the hypothesis was incorrect.

It is this cautious attitude of mind that is necessary in designing experiments and, as we have said previously, all the imagination, experience, and knowledge, that we have is put on test in designing an experiment that will give a clear " Yes " or " No " that is acceptable to other people. I have said that it is worth while repeating other people's experiments if you are going to do them more carefully. In contrast your own work should be done so carefully that it will stand the test of time and be so convincing that no one will feel that it incites repetition.

In carrying out this final test with a six-chambered revolver you must pull the trigger 6, or for extra caution, more than 6, times, watching to see the cylinder rotate on each occasion. To spin the cylinder each time before pulling the trigger—as is done in Russian roulette—or to pull the trigger 5 times and not 6, proves nothing at all. Every chamber must be tested systematically. Applying this more generally, if you have not the money, or the time, or the patience, to carry out the complete experiment then do not bother to start it, do not waste the good money you have got. Do not end up by writing an article full of hypothesis and conclude with a statement that: " I pulled the trigger 3 or 4 times, there was no explosion so I assumed the gun was unloaded." You may convince yourself but you will not convince other people.

Thus a well planned experiment answers the question that is put to it with a clear " Yes " or " No." It is acceptable by other people because they can see that: (i) You have tried all possible contingencies, (ii) you have carried out a trial in which you have put your own animals to the same risk that you are asking them to expose theirs to if they act on the result of the experiment, (iii) that the plan of the experiment is one that will completely destroy the hypothesis if it is false, and (iv) it is a direct experiment using a practical technique and it does not rely on a complicated technical method that could introduce its own errors.

In addition to its primary object a simple experiment can have two other objects: (i) answering subsidiary questions, and (ii) the collection of further data. These additional objects will be dealt with later.

Primary object

The introduction indicates that to make sure the primary question is answered with a clear " Yes " or " No " the problem is not to be tickled, or patted, but hit with a hammer. To illustrate this metaphor, suppose that the hypothesis is that it is dangerous for an animal to eat more than a certain amount of a particular plant. To test this hypothesis you do not give just twice the stipulated amount, but you give just as much of the plant as it can be made to eat, and then mince up more of it and administer it with a stomach pump. This amount should be so large that if it fails to cause any ill-effect, then this part of the investigation is complete, for it would not be practical to try to give a larger amount. But, if the plant is toxic, then the worker will be presented with an acute disease, in which the symptoms are marked and obvious, and in which the post-mortem examination should show clearly which organs are affected.

The reverse of this method is to start at the smallest dose that might be toxic, and then work up if necessary. This method often resembles the antics of an unskilled driver, trying to turn his car in a narrow lane; it may end entirely without result because the experimenter runs out of either animals or material. Further, although it may be done out of kindness, it will often result in causing pain and distress to a larger number of animals than does the cruder method.

Once it has been clearly shown that an extravagant dose is harmless then, and then only, is the time to follow up with other questions of toxicity, such as giving small doses over a long period, or giving the plant in combination with other factors. Similarly, if the plant is toxic, secondary questions will be the extraction of the toxic principle in a pure form, and the quantitative assessment of its toxicity.

The primary object is a fundamental one, and it is not necessarily related to what is likely to occur in practice. Thus, to prove the toxicity of a plant, an animal may be fed with all the specimens of the plant that can be found over an area of many square miles, even although it would be quite impossible for the animal to feed over such an area. Similarly, the treatment of a disease, or a method for increasing carcase weight, may cost, in the preliminary experiments, several times the value of the animal, and the question of economic methods can be left until the circumstances are better understood. It is, however, very desirable, when working with commercial units, to explain this to the animal attendants, for they have their own ideas on scientists, and may feel that to spend more than the value of the whole animal on obtaining a slight increase in carcase weight, is merely another sign of naïvety.

Controls

Controls are the standard by which the result of the experiment is measured. Without a mark to measure distance from, there can be no measurement. Without measurement there can only be impressions, opinions, and arguments, the emotional playground of the crank. The continual cry of the research worker is: " The controls have spoiled the result " for the sick animals make a good recovery, but so do the controls! The mineral-fed animals show a good increase in weight, but so do the controls! Without controls research work would be much more fun, and for this reason investigators who lack training, or self-discipline, tend to avoid them as much as possible, and prefer to compare their experimental animals with memories of other animals, at other times, in other places.

Controls are sometimes classified as either direct controls, or indirect controls, and the ability of an experimenter can be judged by his ability to choose the most direct control. The direct control is based on the method of difference, in which, broadly speaking, there is only one circumstance that is different between the experimental and control group. The term " broadly speaking " is introduced because under practical conditions, it is impossible to have two mice, or even two environments, exactly the same. Indirect controls have more than one circumstance different between the groups, and therefore leave subsidiary questions to be answered: they may be used instead because the experimenter has not the ability to devise a direct control, or to save time and money when experience suggests that a direct control is not essential.

The difference between the two terms is best shown by examples. Beginning with the example used in the previous section, which dealt with dosing an animal with a large amount of a plant that was suspected of being toxic, and recording the effect. Here the *indirect control* would be the recording for a healthy, and similar animal fed on a routine diet. The *direct control* would be to feed a control animal with an equal dose of an innocuous plant closely similar to the one under investigation.

In some diseases the blood serum from an immune animal affords protection to an animal injected with it, a phenomenon known as passive protection. In experiments on passive protection, the *indirect control* would consist of giving the test dose of infective organisms to unprotected control animals, whereas the *direct control* would consist of testing control animals that had received the same amount of blood serum from non-immune animals.

In some circumstances it may be desirable to use both direct and indirect controls, so that more than one question can be answered. Thus in a histological study of lesions caused by a live bacterium injected into the skin, the *direct control* would be the injection, elsewhere, of a similar dose of dead organisms, whereas the additional *indirect control* might be an uninjected piece of skin from a similar part of the same animal. One could go further than this and, if the organism was believed to produce a soluble poison, one could also inject fluid alone, having previously filtered out the organism, thus adding a second indirect control.

More than two groups

We have seen that, even where a single question is being asked, the answer to subsidiary questions may be given by including both a direct and one or more indirect controls. These additional questions are not, however, restricted to subsidiary questions but may be alternatives of equal importance. The more questions that are asked in one trial, the greater the number of animals that are required, but also the greater the proportional saving in animals in that one direct control group can cover all the different treatments.

The disadvantage of using more than one treatment is that the number of suitable animals may be limited, and under these circumstances no additional question may be asked if it will imperil the clarity of the answer. To give an example of this, let us assume that we are going to use about 20 rats in a maze but, before starting anything more elaborate, we wish to see if, as a group, there is any bias towards either turning left or right.

If we divide the end of the passage into two equal openings and run the rats through, we might find that only 4 rats turned left and 16 turned right. We can compare this with the null hypothesis that 50 per cent would go each way by looking the result up in Mainland's binomial tables. Looking up 4 *A*'s in a sample of 20 we see that with a 95 per cent confidence level the true number of *A*'s could vary from 5·75 per cent to 43·65 per cent. This upper limit is less than the 50 per cent of our null hypothesis and therefore there is reason to suspect that the rats have a bias towards turning to the right.

In contrast to the above, suppose that we are not yet trained research workers, and still have a fear that simplicity denotes ignorance. Under this stimulus we try not only to show a bias but to measure its strength. We could try to do this by moving the position of the partition. Thus with a passage 20 cm. wide we may put the partition first at 5 cm. then at 10 cm., and lastly at 15 cm. from the left hand wall. If, for simplicity, we add another rat, then we can run 7 rats through with the partition in each position. Under these conditions, if by any chance there is a strong bias for the rats to make for the larger opening, practically all the evidence we are left with to decide the left or right bias is the result for the 7 rats that were used when the partition was in the middle position. To get a significant result under these circumstances every rat in the group of 7 would have to make the same choice.

This very simple example is explained in rather tedious detail because it does illustrate the very important principle that, if you ask for too much from a fixed number of animals you may end up with no answer at all.

Value of concomitant observations in design

To avoid generalities and to give some specific guidance on the design of a simple experiment let us take as an example a comparison of the milk yields of two groups of cows, one group milked by hand and the other by a milking machine that is under test. We shall then be able to see how modifications in the design can alter the efficiency of the experiment.

Multiple Samples

In this experiment 10 cows were available for each group and 3 weeks was considered to be a satisfactory trial period. The type of data is illustrated in Table 165 which shows the results for 3 cows out of the 10 in each group and the purpose of the Table is to emphasise the danger of multiple samples, a danger that was mentioned earlier when dealing with the sampling of populations and with regressions.

Table 165 gives an indication that in these cows the milk-yield is falling during the period of the experiment. A fall that usually amounts to about 10 per cent per month. Now, if we were going

TABLE 165

Weekly milk yields in lbs. for 3 of the 10 cows milked by hand and 3 of the 10 cows milked by, machine over a period of 3 weeks.

Treatment	Cow	Yield for previous week	Wk. 1	Wk. 2	Wk. 3	Tot.	Average
Hand Milked	BR	263	264	242	248	754	251
	MA	276	271	252	267	790	263
	IG	239	230	220	234	684	228
Machine Milked	HH	266	232	211	200	643	214
	OB	240	216	191	193	600	200
	ES	269	260	254	182	696	232

to calculate this fall and compare the regressions in the one group with the regressions in the other we would require to use all the data in the Table and therefore the total degrees of freedom would be 18—1. But instead of this very complicated comparison of trends we can make a direct comparison between the milk-yields of the individual cows in the one group and those of the individual cows in the other.

If we do this, then we can use either the total or the average milk yield for each cow, and thus, with 6 cows, we will have 5 d.f. We cannot use 18—1 d.f. for under these circumstances the weekly yields are akin to multiple samples.

Having made this point we now go back to considering the complete groups of 10 cows in each.

Simple Groups

The simplest design we could use in this experiment would be to use simple groups and divide the 20 cows into 2 groups by using random numbers; the use of random numbers is explained in a later chapter. We could then toss a coin to decide which group was hand milked, and which milked by machine.

The results for such an experiment are shown in Table 166. with the average yields in lb. rounded off into gallons (10 lb. = 1 gal.) to ease the arithmetic. The method of analysis will be the same as that already used in Table 85.

TABLE 166

Average milk yields in gallons per week over period of 3 weeks.

Hand Milked		Machine Milked	
Item	$Item^2$	Item	$Item^2$
21	441	16	256
25	625	21	441
23	529	20	400
27	729	21	441
26	676	23	529
14	196	13	169
11	121	12	144
8	64	9	81
14	196	12	144
10	100	8	64
179	3677	155	2669

Total $179+155 = 334$

Total for $(\text{items}^2) = 3677+2669 = 6346$

Main Correction Term $\dfrac{334^2}{20} = 5578$

Total s.sq. $\Sigma(\text{Items}^2) - \text{M.C.T.} = 6346-5578 = 768$

S.sq. between groups $\dfrac{(179)^2}{10}+\dfrac{(155)^2}{10} - \text{M.C.T.}$

$= 5607-5578 = 29$

Analysis Source	*S.sq.*	*d.f.*	*V*	*F*
Total	768	19		
Between groups	29	1	29	0·7
Residue (error)	739	18	41	

It can be seen from Table 166 that the total sum of squares amounts to 768 whereas the difference between groups accounts for only 29, leaving 739 for error. With 20 cows there are 19 degrees of freedom (d.f.) and between the two groups there is 1 d.f. leaving 18 for error. Hence the *F* value is 29 divided by 41 and amounts to less than 1, a long way from significance. Is this lack of significance due to the large s.sq. due to error or is it proof that there is no real difference between the groups?

When we look at the results and see that the yields range between 8 and 27 gallons in one group and 8 and 23 in the other, variations of about 300 per cent, it seems logical that these figures should be unsuitable for showing a difference between two groups with averages as close as 15·5 and 17·9, a difference of under 16 per cent. We obviously need a more sensitive test and we will take an important step to refinement by using randomised blocks.

Note that Table 166 provides us with the total sum of squares, the main correction term and the sum of squares for treatment. These will not be altered by refinements and their calculation will not be referred to again.

Randomised Blocks

The important modification of introducing a restraint into the design changes it from one of "Groups" to one of "Randomised Blocks." The restraint, or restriction, that we will apply is to place the cows into pairs according to their previous milk yields, so that each pair with similar milk yields form a block. By tossing a coin, or by using random numbers, we then decide which member of the pair will be hand milked.

Working on these lines then, in Table 165, the pre-experimental yield of 276 would be paired with 269, 266 with 263 and 240 with 239. One of each pair would then be allotted to its group by tossing.

This design then gives us Table 167 which repeats the yields already given but introduces a column for the sum of each pair, or block, and a column for these totals squared. The Table shows these columns in sets of 5 yields and for the moment we will assume that this has been done to help cross-check the arithmetic between columns and rows.

TABLE 167

Average milk yields in gallons over period of 3 weeks.

	Hand Milked	Machine Milked	Blocks	Blocks^2
	21	16	37	1369
	25	21	46	2116
	23	20	43	1849
	27	21	48	2304
	26	23	49	2401
Sub-total	122	101	223	10039
	14	13	27	729
	11	12	23	529
	8	9	17	289
	14	12	26	676
	10	8	18	324
	57	54	111	2547
Treatments	179	155	Blocks 334	12586

S.sq. for blocks $\frac{1}{2}(1369+2116 \ldots . 676+324) - \text{M.C.T.}$
$= 6293 - 5578 = 715$

The method is now the same as that used for paired data in Table 93 where there were two weighings of the same individuals, except that an alternative arithmetical method was used for finding the sum of squares between blocks. In Table 167 it can be seen that the s.sq. between blocks gives the high total of 715, and as there were 10 blocks, this total is associated with 9 d.f. so that each d.f. will remove 79 from the residue for error, and the analysis of variance will be as follows:

Source	*S.sq.*	*d.f.*	*V.*	*F.*
Total	768	19		
Between treatments	29	1	29	10·8
Between Blocks	715	9	79	29·8
	24	9	2·6	

This analysis shows how the reduction in the s.sq. for error has boosted the F value for the treatment. Using Table V at the back of the book with $n_1=1$ and $n_2=9$ we see that $p=0{\cdot}05$ has a value of 5·1, so that 10·8 is easily significant and, in fact, it lies near $p=0{\cdot}01$ showing that the slightly lower milk yields in machine milked cows are unlikely to be due to chance alone.

The very high F value for between blocks shows that there is more than a chance variation between pairs of cows, an unnecessary piece of information but of some value in indicating the efficiency of the restraint in removing a large sum of squares from the residue.

Randomised blocks are one of the most usual, and most useful, designs for experiments on animals and this design can be used as a safeguard to remove possible bias due to using animals of different age, weight, breed or sex. Each block must contain the same number of animals as there are treatments, or multiples of the number. Thus with 5 treatments each block might contain 5, 10 or 15 animals and the 5 treatments would be randomised among them.

Factorial and Interaction

Whether the reader takes this section as an encouragement to adopt an even more efficient design, or whether he takes it as an awful warning to keep to simple methods, he should, nevertheless, read it so that he can understand the meaning of the word interaction, which he will come across in reading the scientific literature.

Using randomised blocks we were able to show that the milking machine was less efficient than hand milking. By using a different restraint we can test to see if there was any real difference in the reaction of cows with high yields to that of cows with low yields.

In this design there are two alternatives. In step 1 we will use one simple restraint that will block 10 cows as high yielders and the other 10 as low yielders. In step 2 we will do the same but in addition we will block each pair according to similar pre-experimental yields, as was done in the section on Randomised Blocks.

Step 1

Dealing with Step 1 we select 10 high yielding cows for one block and 10 low yielders for the other. We then randomise the cows in each block to distribute them to the two different treatments. This gives 4 sub-groups each with 5 cows, and these groups are shown at the bottom left-hand corner of Fig. 170 (F) and as we have the same number of replicates in each sub-group this design can be treated as a factorial experiment.

A factorial experiment is one in which more than one factor is being investigated, and in which all possible combinations of the factors are used. In this example we can call "machine milked" one factor and "high yield" another. Fig. 170 (F) shows that we have all the combinations. Both factors are present in the top-right sub-group, high yield alone in the top-left, machine milked alone in the bottom-right, and neither factor is present in the bottom-left. Because it has 2 factors this is known as a 2 by 2 factorial (present or absent × present or absent) and it has one interaction. Let us see what this interaction is: but first we must turn back to Table 167.

Looking at Table 167 we can see that to treat the data as a factorial all we need do is to label the top 5 rows "High Yields" and the bottom 5 rows "Low Yields" which gives four sub-groups with sub-totals of 122, 101, 57 and 54. To ease the arithmetic we can code these sub-totals and round them off to give the following 4 cell table.

High Yield	12	10	22
Low Yield	6	5	11
	18	15	33

The total s.sq. which cannot be exceeded by these 4 sub-groups is $(12^2+10^2+6^2+5^2)-\dfrac{33^2}{4}=305-272\frac{1}{4}=32\frac{3}{4}$

The s.sq. between treatments is

$$\left(\frac{18^2}{2}+\frac{15^2}{2}\right)-272\tfrac{1}{4} = 274\tfrac{1}{2}-272\tfrac{1}{4} = 2\tfrac{1}{4}$$

The s.sq. between yields is

$$\left(\frac{22^2}{2}+\frac{11^2}{2}\right)-272\tfrac{1}{4} = 302\tfrac{1}{2}-272\tfrac{1}{4} = 30\tfrac{1}{4}$$

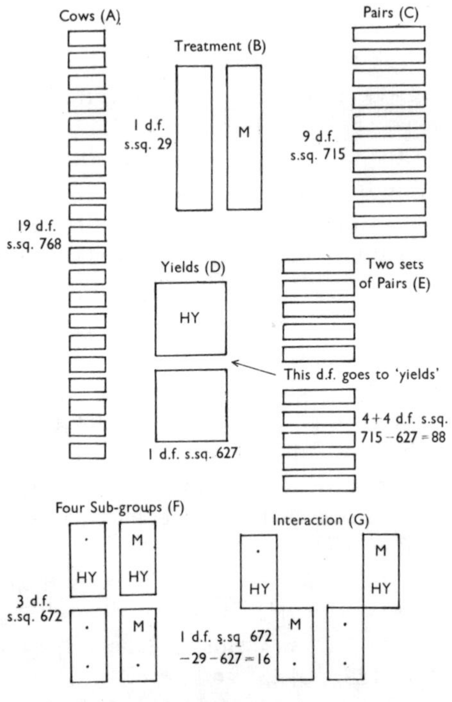

FIG. 170. Degrees of freedom and sums of squares in the milk-yield experiment.

The only other comparison we can make between these four sub-groups is to take them cross-ways, 12+5 and 10+6, giving

$$\left(\frac{17^2}{2}+\frac{14^2}{2}\right)-272\tfrac{1}{4} = 272\tfrac{1}{2}-272\tfrac{1}{4} = \tfrac{1}{4}$$

The s.sq. depending on the cross-over is that called the interaction and it will be seen that the s.sq. for the 3 comparisons, treatment, yield and interaction, which are shown diagrammatically as B, D and G in Fig. 170 add up to the total for the s.sq. for the 4 sub-groups. Hence we can find the s.sq. for the interaction by subtracting the s.sq. for treatments and the s.sq. for yields from the total for the 4 sub-groups, and this is the usual way of doing it.

We can now apply this method to the example in which there are 5 replicates in each sub-group, and as a basis I have repeated the sub-totals from Table 167 in the Table below.

	Hand Milked	*Machine Milked*	*Yield Sub-Totals*
High Yields	122	101	223
Low Yields	57	54	111
Treatment sub-totals	179	155	

Already known from Table 166. Total s.sq. 768 MCT 5578
Treatment s.sq. 29

S.sq. between sub-groups $\left(\frac{122^2}{5}+\frac{101^2}{5}+\frac{57^2}{5}+\frac{54^2}{5}\right)-\text{MCT}$

$= 6250—5578 \qquad = 672$

S.sq. between yields $\left(\frac{223^2}{10}+\frac{111^2}{10}\right)-\text{MCT}$

$= 6205—5578 \qquad = 627$

S.sq. for interaction S.sq. sub-grp.—yield—treatment

$= 672—627—29 \qquad = 16$

We can now examine the result for Step 1 and this is recorded in the following analysis.

Source	*S.sq.*	*d.f.*	*V.*	*F.*	
Total	768	19			
Treatment	29	1	29	4·83	
Yields	627	1			$p = 0{\cdot}05$
Interaction	16	1	16	2·66	where
					$F = 4{\cdot}49$
Residue	96	16	6		

It can be seen from the above analysis that the result for Step 1 is that the treatment is significant at $p = 0{\cdot}05$, (although when we used pairs as blocks $p = 0{\cdot}01$) and that the interaction was not significant. We are therefore in a somewhat worse position than when we used pairs as blocks, and if we wish to get a really sensitive test we must reintroduce them. This we do in Step 2 which follows.

Degrees of Freedom, Step 2

We have seen that for the 4 sub-groups shown in Fig. 170 (F) there was a s.sq. of 672, and as there are four blocks there are 3 d.f. Neither of these totals can be increased. We found that these 4 blocks could be compared as treatments, as yields and as interaction, and that each comparison took 1 d.f. and because this accounted for all the d.f. then the s.sq. for the three comparisons had to equal the total for the 4 blocks, and amount to 672.

Looking now at Cow Yields (A) in Fig. 170 we see the column of 20 yields giving a total s.sq. of 768 and 19 d.f., neither of which can be exceeded. When dealing with *Simple Groups* we split the 20 yields into 2 treatments, with the result that there was 1 d.f. for this comparison, leaving 18 d.f., that is, 9 d.f. in the top 10 and 9 d.f. in the bottom 10.

With these examples in mind we can turn to Pairs (C) which shows 10 pairs, which we found in the section on *Randomised Blocks* to give a total s.sq. of 715 and to have 9 d.f. We can see from the figure at (E) that the comparison of yields can be looked upon as a comparison of the top 5 pairs with the bottom 5 pairs. Hence for the comparison of High Yields with Low Yields we have 1 d.f. leaving 8, 4 within the top 5 pairs and 4 within the bottom 5 pairs. We have found that this comparison of high and low gave a sum of squares of 627 and took 1 d.f. Hence if this comparison is part of the total for pairs, then 715—627, or 88, remains for that left within the 5 pairs in the high yields plus that within the 5 pairs in the low yields.

We can now undertake Step 2 in the analysis, using a s.sq. of 627 between yields but reducing the s.sq. between pairs to 88 with 8 d.f. instead of the original 9.

STEP 2 ANALYSIS

Source	*S. sq.*	*d.f.*	*V.*	*F.*	*t.*	*p.*
Total	768	19				
Treatment	29	1	29	29	5·4	<0·001
Yield	627	1				
Interaction (672–29–627)	16	1	16	16	4·0	0·004
Pairs (715–627)	88	8				
	—	—				
Residue (Error)	8	8	1			

This example emphasises how important it is to make use of concomitant observations, or supplementary observations, when planning an experiment, and how much more sensitive the experiment becomes when they are used correctly. The same applies to restraints used under the heading of stratification when making surveys.

For easy comparison the effect of the different restraints demonstrated in this example are summarised in the following Table.

	Simple Groups	Randomised Blocks (Pairs)	Factorial (Yields)	Factorial (Yields) (Pairs)
Treatment	Not sig.	0·01	0·042	0·001
Interaction	Untested	Untested	Not sig.	0·004

Looking again at Fig. 170 (G) we can see that in testing the interaction we had 5 cows with high yields and 5 machine milked on the one side and 5 with high yields and 5 machine milked on the other, and that we obtained a highly significant result. What is the meaning of this highly significant interaction? Some authors appear to be completely overawed by it and would record that "the interaction of yield and treatment was highly significant" almost as though the statistician had handed them an unexploded bomb that they knew neither how to use nor how to get rid of.

The meaning of this interaction is found in the 4 sub-totals for the sub-groups, but to make it clear it is often an advantage to put the results into a figure as I have done in Fig. 174. From this figure it can be seen that the lines for high and low yields are not parallel. The fact that the interaction is highly significant shows that the lack of parallelism is not likely to be due to chance. Hence we can say from the test of treatment that the machine tested was less efficient than hand milking, and from the interaction, that this lack of efficiency was more pronounced in cows giving high yields.

This positive interaction might pose another question. Does it indicate that the efficiency of the machine is in some way proportional to the milk-yield?

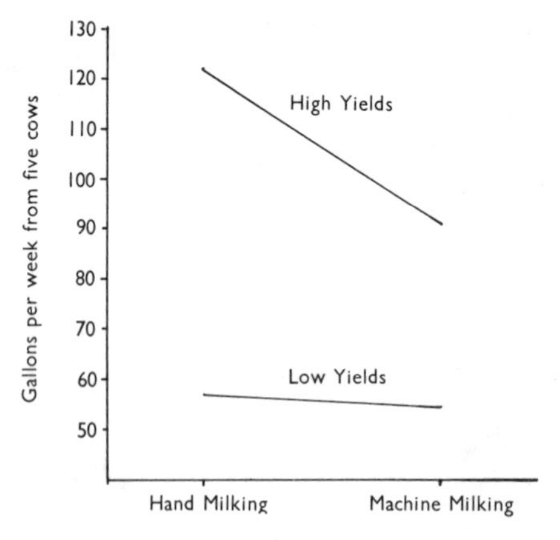

FIG. **174.** Results for experiment on milk-yields recorded as a diagram.

Note that, when shown as a diagram the interaction lines may converge, diverge, or even cross, but where they appear to be parallel it would not be worth calculating the interaction. With highly complicated factorials there are second, or even third order interactions, but these are outside the scope of this book.

Use of Restraints

Note that the way in which we made the analysis of the milk yield experiment more sensitive was by using concomitant observations, which in this instance were the milk yields prior to the experiment. Previously, in the chapter on regressions, we made use of concomitant observations by working out an index of response, and suggested that more experienced workers could use an analysis of covariance. We could have used either of these methods here. Thus we could have worked out an index of response using the average yield of each cow during the experiment as a percentage of a similar period prior to the experiment. Using this method we would, as before, have to make assumptions that the data was based on straight line regressions and that the difference between groups was one of a simple addition.

In the experiment on milk yields, our supplementary observations, although still called concomitant observations, were available before the experiment began, and hence we could avoid both the need to make assumptions, and the calculation of an index of response, by using the pre-experimental yields as a basis for placing restraints.

The test might have been slightly more sensitive if we had used an analysis of covariance in addition to placing restraints, but without this we were able to show that the result was unlikely to be due to chance alone.

Comments on Factorial Design

It can be seen that a factorial design can be a very sensitive one and provide information, not only on the treatment, but also on the interaction, or, in more complicated designs, the interactions.

As an example of a more complicated design take an exploratory experiment on a vaccine. In this there could be a comparison between three different doses of vaccine. To keep the vaccine in the tissue they might be used in conjunction with three different additions of adjuvant, and the challenge, to test immunity, might be a small test dose to one half and a large test dose to the other. This would give a $3 \times 3 \times 2$ factorial with 18 cells, or plots. If the result was to be assessed by a measurement, say body weight, then as the sensitivity of the experiment would depend on the reliability of the average for each cell, one would be advised to have at least 5 animals in each cell, making a total of 18×5, or 90 animals. Alternatively, if the result was to be judged by a qualitative criterion, "sick" or "well", "deaths" or "lives", instead of measurements, an even larger number would be used.

What are the disadvantages of these more complicated designs? Firstly one must remember a warning given by Cochrane and Cox (1953) that the most common faults in design are vagueness and excessive ambition, through which a worker may try to do 20 years' work in one experiment.

If one rushes into a complicated design without adequate preliminary work many snags are possible. The selected doses may be too near each other to measure the difference in result. If the effectivity of the vaccine was based on a faulty experiment the only result of a subsequent complicated and expensive factorial design might be to show the vaccine useless. A test organism used to challenge immunity might have lost its power since it was last measured and thus leave the experimenter with 90 health animals with unknown differences in resistance.

Another danger is that the experiment will be confounded; this is dealt with more fully in the next section. Thus if, say, the animals in 3 cells fall sick with an intercurrent infection, the whole balance of the experiment may be upset and the additional sensitivity of a factorial design will be lost. Admittedly there is a method of calculating a theoretical value for a missing cell but obviously this method has its limitations.

Finally for a factorial there must be an equal number of animals in each cell. Thus for 18 cells of 5 we need 90 similar animals, or if we cannot find these, there must be 5 groups of 18 similar animals, and it is often difficult to find these. Further, not only should we start with 90 live animals but our experiment may demand that we end with them, and if the experimental treatment kills some of them then we are left with the problem of predicting an appropriate measurement for that cell.

Thus until experience has given one the knowledge to judge for oneself it is wiser to restrict oneself to randomised blocks and the simpler forms of factorial design, even if this means carrying out 3 or 4 smaller experiments rather than one elaborate one, for under these circumstances one can use one experiment to judge whether the dose should be bigger, or smaller, in the next one. The additional safety is paid for by the fact that each step must have its own control.

CONFOUNDING

The ideal for an experiment is that the only difference between the experimental group and its direct control is one factor, and this factor usually represents the experimental interference. If there is more than one difference between the two groups, and this difference is a possible source of bias, then the experiment is confounded.

Confounding can be intentional or accidental and unrecognised. Intentional confounding can be introduced to save time by trying " shotgun " treatments, in which several remedies that may have been advocated can, provided that they are unlikely to be antagonistic to each other, be tried together. If the experimental group shows no advantage, then all the remedies are discounted and time is saved. Alternatively, if the experimental group shows a worthwhile advantage, the various remedies can be tried out in an expensive factorial experiment, testing both the treatments and their interactions, safe in the knowledge that one treatment, or interaction, should show an advantage. Intentional confounding may also occur in elaborate factorial designs, and here the second, or some higher order, interaction may be sacrificed by confounding in order to allow a fuller judgement on the more simple treatments, but this is for experts.

A simple ignorance of the existence of confounding is shown in those technical papers in which cows of one breed, in one environment and given a commercial supplementary ration, are measured, and these measurements are compared with cows of another breed, in another environment, receiving supplementary rations of another make. In these circumstances an author will often claim breed differences without realising that he would be equally justified in writing a second paper recording the same changes as the effect of diet, or a third paper giving the changes as the effect of the good hygiene and the healthy environment of one herd.

Even under more controlled conditions, confounding can occur unwittingly. Thus one animal infested with parasites, or some other infectious complaint, may be allocated at random to a group, and either contaminate most of its companions or leave an infected environment for the next occupants. Other factors that may confound an experiment are one cage of mice being exposed to a draught from which the other cages are protected or one group of animals housed in a building that catches the mid-day sun; alternatively, one group may be housed in a building in which the milking machine, or weighing machine, or some other device, is faulty. A more subtle difference may lie in the stockman, and a really good stockman may confound every experiment because his cattle always show a better condition of health than has been expected.

In many experiments these factors can be controlled by intermixing the control and experimental groups but where this is not possible, for example, when working with infectious diseases, then the experiment must include a sufficient number of replicates to allow each group to occupy each environment in turn, and this is an additional reason for simplicity in design.

Single and multiple experiments

In general, one big experiment will produce a more significant answer than two small ones. Snedecor (1950) gives the simple illustration of a man handed 3 samples of beer, one of which has received an experiment treatment.

The probability of him picking out the treated sample by chance alone would be 1 in 3. If the experiment was repeated the odds would be $\frac{1}{3} \times \frac{1}{3}$ or 1 in 9. In contrast, if the 2 experiments had been combined and the man had been given 6 glasses of beer, of which 2 had been treated, then the possibility of him picking out the correct glasses by chance alone would be $\frac{2}{6}$ for the first glass, and $\frac{1}{5}$ for the second glass, giving a chance of only 1 in 15. Thus the single large experiment has produced more convincing odds. Even with this simple experiment there could, however, be a snag. If each sample represented a pint glass then Mr. Smith might pick out the first treated sample at the first sitting, and the second treated sample at a second sitting. But if presented with the 6 pints at once, Mr. Smith might himself become confounded. This would be an example of the circumstances changing during the course of the experiment. Under some circumstances, however, the interest may not lie in the ability of Mr. Smith, but the question might be: can anyone pick out the treated sample? Here it is obviously more useful to have the small test on Mr. Smith and a similar test on someone else.

Thus as a general rule, where you want the highest level of significance you use all your animals in one large experiment always providing that you know that the conditions will not change during the experiment; alternatively, where you wish to know that you can repeat a favourable result at will, you carry out the same small experiment on different animals, at different times and under different chance circumstances.

Complicated experiments and statistical help

Complicated experiments involving confounding are outside the scope of this book, but readers who have been told that they have designed a split plot experiment may wish to know what this means. In that case they will find a simple logical explanation of this method in animal experimentation, given by Kempthorne (1952) in example 19·6.

If an elaborate experiment is contemplated then advice should be obtained from another worker who is working on the same lines, and who has successfully dealt with similar problems; or from a statistician who is used to working on biological problems. Be cautious of those who have a theoretical knowledge of statistical methods but lack the practical experience necessary for their digestion, their tendency is towards over elaborate experiments.

Green (1954) pointed out that, if asked for advice, the role of the statistician should be " that of obstetrician, rather than morbid anatomist, for it is unfair to expect him to extract scientific knowledge by performing a kind of mathematical post-mortem upon the numerical remains of a badly planned study." A criticism on this statement, given by another statistician, was that he thought it would be even better if the statistician was allowed to share in the pleasure of the conception, instead of coming in at the birth.

Working with a statistician is a great advantage, providing that both the experimenter and the statistician are sufficiently sure of themselves to be willing to explain the simple logic that lies behind their complicated jargon. It is not easy when both are frightened of appearing stupid, and regard their vocation as a priesthood that must be defended; under these circumstances each regards the other's questions as an impertinent prying into the efficiency of his own methods.

Once a good partnership has been set up, then any vagueness in design will disappear with discussion, and the result of the experiment will be easily appreciated by both workers. Where two workers have worked on the design of an experiment then both names should appear in the published results; here the experimenter must realise that, in later years, the fact that all his experiments were workmanlike will give him far more satisfaction than the fact that his name stands alone on a few publications of less than mediocre value.

Collaboration with cranks

We have said that some workers have investigations thrust upon them, and sometimes this is done through the persistence of some individual who on some subject or other is more than two standard deviations from the average concept, and who, to keep things concise, may be designated a crank.

The main attributes of a crank are his powers of persuasion, his unshakeable belief in his own hypothesis, his knowledge of all the literature and incidents that can be made to support his hypothesis, his imagination, enthusiasm and apparent sincerity. He eliminates data that do not support his hypothesis by disbelief or by the introduction of another hypothesis, and he is quite incapable of designing an experiment that could show his hypothesis to be untrue.

An experiment with, or for, a crank, should not be undertaken lightly, but as it is only an exaggerated instance of collaborating with any other person it is worth a little discussion. One reason for caution is that, should the result of the first experiment be unfavourable, a second hypothesis will be presented to the effect that " These animals were too old " or the " Test dose was too high," or if the measurements were carried out by the crank himself, the excuse may be made that this was the first time he had used his apparatus under such conditions, and that owing to a maladjustment the rays had gone through the wrong hole, or that the samples were on the small side as cytoplasmic oscillations varied within the material and with a larger sample this variation might cancel out.

The first action is to examine his evidence and to cut through what are called evaluative words (such as " very effective ", " opinion of eminent scientists ", " repeated successful experiments ") which merely state opinion and create mental atmosphere, and search for the descriptive words which express fact. When you understand the meaning of these statements then you must think out a method by which they can be verified, and then plan an experiment that will do this.

In trying to design such an experiment you may meet resistance, for cranks prefer experiments that do not challenge the main theory. To illustrate this point; if he says that he is sure his treatment is effective on rabbits, he will want you to try it on, say, sheep. If you insist on rabbits he may be indignant that you should cast doubt on his integrity and accuse you of trying to waste both your own time and his. What he wants to know is; does it work on other animals, such as sheep? If you permit an experiment on sheep and get a negative result then all he may say is " What a pity, well now let's see if it will work if we use pigs." This is time consuming, for you can go from pigs to cows and to poultry without ever showing his original statement to be wrong, while any result which favours the hypothesis in any degree will be quoted *ad nauseum* even if further experiments show that it was fortuitous.

In designing an experiment then, besides ensuring that it can if necessary show his hypothesis to be wrong, it is very advisable to make him commit himself on what would be the poorest result that he would consider sustained his hypothesis. Thus having shown him the plan of the experiment, the type of animals to be used, the size of the dose, and the method of measuring to be used, it might be possible to obtain a written statement that he can guarantee that one group will end up at least 20 per cent heavier than the other at the end of 3 weeks. By this method a chance result producing a trivial advantage to the treated animals fails to sustain the hypothesis.

Cranks usually feel that the routine experimental methods are especially designed to trap them and will try to confound the experiment by introducing some other factor in the form of a secondary question, which will give them a way out. If a true experiment cannot be mutually agreed upon it is best to report the proposed experiment and say that your associate cannot agree to it. Where you carry out a true experiment and he proves his point, well and good, you can report the result. Alternatively, where the result does not sustain the hypothesis then confine the result to some statement that the experiment mutually agreed upon was carried out and the result did not justify the claim made by your associate. Be sure that if you put too much detail into your report he will find something to use as a lever to justify another experiment.

Summary

The experiment is the main tool of the research worker. It can be divided into three main parts: 1. The primary question; 2. Secondary questions; and 3. Collecting additional data. The primary question is devised to test the hypothesis and prove it to be wrong, if it is wrong. The answer must be a clear " Yes " or " No," so convincing that there will be no suggestion that the experiment needs repeating. Secondary questions may be asked providing they in no way imperil the primary question, and this usually means that the experiment should have a simple factorial design. The collection of additional data will be discussed in a later chapter.

When carrying out an experiment in collaboration with other people make no exception to these rules, and insist that the experiment must be capable of showing the hypothesis to be wrong. If suspicious of a collaborator come to a clear understanding of what will be the poorest result that can be held to sustain the hypothesis, and do this before the experiment starts.

A partnership between an experimenter and a statistician in which one, or both, insist on remaining incomprehensible to the other, because they will not, or lack the ability to, explain the simple logic behind their technique, cannot be expected to be worthwhile. In contrast, a happy partnership produces a very strong combination.

References

Cochran, W. G., and Cox, Gertrude M. (1953). " Experimental Designs." Wiley and Sons (New York). Section 1.22.

Green, F. H. K. (1954). " The Clinical Evaluation of Remedies." *Lancet*, **2**, 1085.

Kempthorne, O. (1952). " The Design and Analysis of Experiments." Chapman and Hall (London). Expl. 19.6.

Snedecor, G. W. (1950). " The Place of Statistical Methods in Biological and Chemical Experimentation." A symposium. Annals of the New York Academy of Science, **52**, 792.

Chapter 10

FOUNDATIONS OF AN EXPERIMENT

Figures with no indication of experimental error may carry unjustified weight.

Rosemary Briggs

In a previous chapter it was said that the reliable research worker is known by the foundations he digs, even though they may be hidden when the building is erected. This preliminary work includes the collection and storage of sufficient material; selection, or breeding of suitable animals; acquiring and testing suitable instruments or techniques; a knowledge of the reliability and errors in the methods; finding suitable test doses; and arranging to start at the proper time of the year. It is convenient to treat the question of error first.

Error

If at great personal discomfort you tried to run a mile in five minutes, and at the end of that distance you were told by your time-keeper that he did not know if you had succeeded or not because his watch did not seem to be reliable, you might feel aggrieved. Similar incidents may happen in research work, sometimes even without the reliability of the watch being questioned. This is partly because some people get the impression that a university training has made them free from error, and to suggest that their method has error is to suggest that they have not been properly trained; additionally they may feel that if an expensive instrument registers 7·853, that must be the correct answer because it is plainly registered on the dial. Nevertheless, there is error in all techniques, and we should be aware of it.

Systematic Errors

These are errors introduced in the manufacture of instruments, or in the preparation of materials. They may be absolute, that is a fixed amount too much or too little, or relative, that is a large amount will show a greater error than a small amount.

If we take a measuring cylinder as an example then we are making use of the fact that, providing the bore of the tube is the same all the way up, the length of tube filled and the volume it contains are proportional, so that we can fit a scale in which double the length equals double the volume.

To demonstrate an absolute error we can consider the bottom of the cylinder where it becomes narrowed and cup shaped through being sealed off. To make allowance for this alteration in bore the maker can put in a measured amount of fluid to fill up this narrow area (say 0·1 ml.) and then start the scale at 0·1 ml. instead of zero. If this small volume at the bottom was miscalculated by putting in 0·09 ml. instead of 0·1, then that would constitute an absolute error, and assuming that the rest of the scale was correct then whatever volume was measured it always be 0·01 ml. short. Thus an apparent 10 ml. would be 9·99 ml. and an apparent 50 ml. would be 49·99 ml. With an absolute error one must therefore add or subtract the same correction factor whatever the amount measured.

To demonstrate a percentage systematic error, suppose that the bore of the cylinder was slightly narrower than had been estimated, then 10·1 ml. on the scale might record an actual volume of 10·0 ml. In this case 50 ml. would be recorded as 50·5 ml. This would be a percentage error and here the correction would be carried out by multiplying the result by a correction factor of $\frac{100}{101}$.

This kind of systematic error is also found in chemical reagents and with these reagents, when they have been made up according to a complicated formula and yet found to contain a slight error, it is usual to write out a correction factor on the label rather than try to alter the proportions of the ingredients.

Complicated Errors in Scales

The acceptance by many individuals of the belief that every manufactured instrument is completely accurate is a compliment to the integrity of the scientific instrument maker, but enough has been said to show that accuracy is a comparative term, yet in addition to the errors already mentioned must be added error with regard to the scale. Keeping to our comparison with a measuring cylinder, this fault would occur if the bore was not of equal diameter all the way up, it could taper one way or the other, or even worse, it could be irregular with one or two unnoticed constrictions. Because of this type of error some manufacturers, in the more complicated instruments, guarantee a certain accuracy over only part of the complete scale.

An error in the scale can cause chaos in an experiment and to illustrate this let us imagine that we are working out the relation of growth rate and diet, and perhaps even weighing out both food and animals on the same balance. After a nice growth curve has developed it slowly veers away from its original course. The experiment is repeated and at about the same weights the curves veer again. This can give rise to a host of fascinating hypotheses, particularly if there are older records to compare it to but, until somebody suggests a systematic test of the weighing machine throughout its scale, the true cause may remain hidden. Thus in spite of our faith in manufacturers, or in our own workmanship, it is best to test an instrument by some simple arithmetical series of samples that will extend over the whole scale of the instrument before using it in an experiment.

Personal Error

A careless, or untrained, worker can show a greater accidental error, which will be dealt with later, but personal error usually refers to a bias shown by an individual. Thus one person likes to make sure that he gives enough, whereas another is frightened of giving too much. An important personal error is found in estimations of colour. This is demonstrated in Fig. 186 which gives a histogram recording the average figure given by each of 12 individuals after each had estimated the same series of 5 dilutions of a brown fluid by means of a graduated colour disc. It can be seen that one made the average 7·48 g/100 ml. compared with another that recorded 8·56, a difference of some 12 per cent. The difficulty in such circumstances is to know which of the 12 observers is correct, for the senior officer cannot thereby claim freedom from error. Here the decision must be reached by testing the material by another method that does not involve judging colour.

The second figure records the estimations on a series of 6 dilutions made by two observers, and it suggests that once you know which is correct, the correction factor for the personal error of the other may be a simple addition, or subtraction, of a fixed amount.

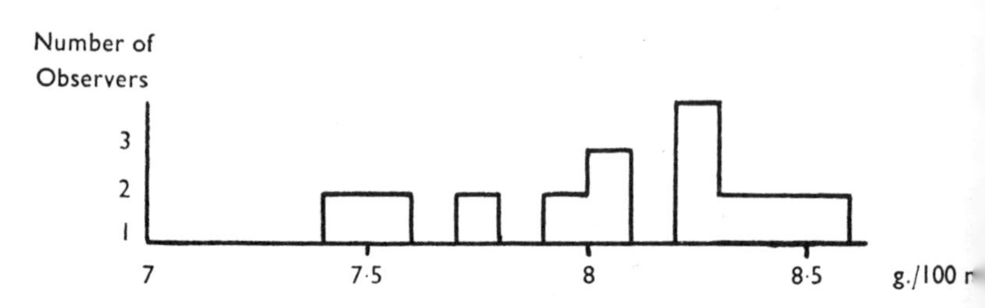

(*a*) Histograms of averages for each of twelve observers making estimates of five dilutions in an arithmetical progression.

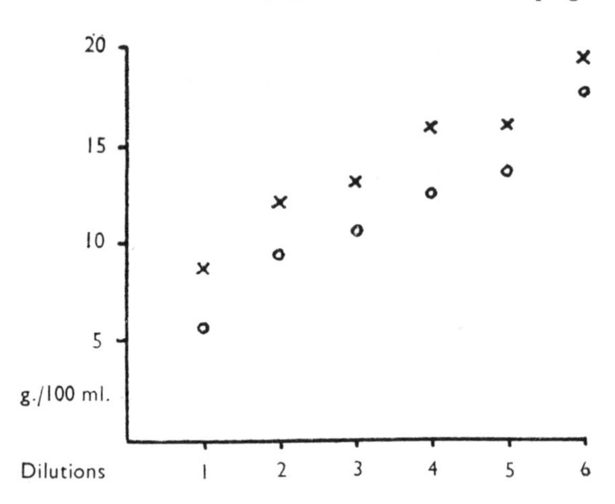

(*b*) Readings recorded for six dilutions in an arithmetical progression by two observers (X and O).

FIG. 186. Alkaline haematin estimated by colour comparison.

Accidental Error

It has been shown that both systematic and personal error can be compensated for by using a correction factor, but one type of error still remains, this is random error.

Accidental error is the sum of all the small errors that in themselves are too small to be noticeable. Thus, using a simple bulb pipette, the meniscus may appear level with the mark at the top when viewed with the naked eye, but a magnifying glass would show that often it was a little too high, or a little too low. To discharge the pipette it should be held nearly upright and variations in this slope can make a slight difference; it should be allowed to discharge for a fixed period and held so that the tip just touches the side of the vessel that is to be loaded, and additional errors can be made in these operations. As most tests require more than one pipetting, together with the mixing of two or more fluids, there is usually room for many tiny errors. Some of these errors may be positive and others negative, so they resemble in effect the passage of shot down Pearson's wedges, and thus give the familiar cocked-hat distribution.

This type of error cannot be removed but it can be restricted, firstly, by skilful and careful manipulations and, secondly, by repeating the test to obtain the mean of two or more readings. As the distribution of this error gives a cocked-hat distribution, the range can be measured by the standard deviation, so that the range of error in duplicate readings will be the s.d. of the range for an individual reading divided by the square root of 2, that is $\frac{\text{s.d.}}{\sqrt{2}}$; thus, if one had time to do each test 4 times the error would be $\frac{\text{s.d.}}{\sqrt{4}}$ in comparison with a single reading, and in consequence the error would be halved.

Dealing with Errors

Systematic error may not be important, in that the heavier object will always register the higher weight whatever the systematic error of the weighing-machine. If, however, the correlation between weight and volume was being investigated, then a percentage error in one measurement would make lines, that should be parallel, diverge. Systematic error is measured by estimating a series of items the different values of which have already been determined by a more accurate, and probably more elaborate technique. Once estimated, the error can be dealt with by a correction factor.

In the same way personal error may be unimportant if only relative estimates are required. Even if some estimates are made by another person his recordings can be corrected to those of the first person by a correction factor, without worrying which individual is nearer the true value. As with systematic error, the correct values are found by using a correction factor based on known standards.

Accidental error, although restricted by skill and care, cannot be eliminated. The standard deviation of its range can be measured by repeated estimations of the same sample, or, preferably, on samples of high, low and medium values. Alternatively, the range can be estimated by making duplicate estimations over a series of actual experimental samples. With this type of error no correction factor is possible, but the result can be followed by the s.d., or the coefficient of variation, so that those interested can compute 95 per cent, or 99 per cent confidence levels.

Test Doses

When an elaborate experiment lasting, perhaps, several months has been designed, nothing can be more embarrassing than to apply the test dose and then find that either it kills all the animals of both experimental and control groups, or that it leaves both groups unaffected. Because of this, finding a useful test dose, when one is required, is an important part of digging the foundation.

Finding a test dose is rather like ranging a field gun by observation. The first object here is to estimate the range and then to try to drop a shell beyond the target. If the first shot does go beyond the target, the next shot is sighted for 300 yards less, on the chance that this will fall short of the target; if it does not, then another 300 yards will be dropped. Once in a bracket of 300 yards, where one shot is beyond the target and the other falls short, we can, by interpolating two more shots differing by 100 yards, limit the bracket to distances of 100 yards. One practical point is that if, when ranging, you think the dust from a shell is beyond the target, when in fact it is in front, you can waste an awful lot of shells creeping up to the target. In the same way, when finding an effective dose, you can waste an awful lot of animals if you base your series on a single trial in which all the animals became affected. Hence, in general, the first objective is to make sure of a dose that will be 100 per cent effective.

Even when the substance is a therapeutic one, and it is believed to be benign, a similar expedient should be used, for at a later date it may be most important to know the possible side-effects of a dose many times the size of the therapeutic dose.

Having found a dose that affects every animal, one must seek a dose that affects none of them, and then seek for our target between the two. In this we depart from the artillery ranging technique, for that is based on an arithmetical progression, involving simple addition and subtractions, which is only suitable for small changes, whereas nature appears to work in geometrical progressions. As an example of this in physiology we have Weber's Law which is based on the fact that the smallest perceptible change in, say, weight or colour, works out as a definite percentage increase on the previous amount. As an instance of this we can say that the increased stimulus of adding a pound weight is immediately obvious to someone already holding 1 or 2 pounds, but adding a pound weight to a load of 60 Lb. may not be discernible, and an increase of, say, 10 per cent of the load already carried may be necessary to produce a noticeable increase. Similar laws apply to dosing with toxic substances, and therefore scales of doses are, in general, logarithmic.

Quantum response

It has been said that, in general, the stimulus of a dose increases logarithmically, but this increase in stimulus in the individual animal is often difficult, or impossible to measure. Under these circumstances the dose must be given to a group of animals which can then be classified as " affected " or " non-affected," and the proportion that are affected can be noted, this is known as a quantum response.

Although the reaction of an individual animal may be recorded as a straight line where the dose is increased as a geometrical progression, this does not apply to a group of animals, and here again we find the cocked-hat distribution, for with most noxious substances, some animals appear to be particularly susceptible, some appear particularly resistant, and the mode is somewhere between the two.

Probits

If the distribution of affected animals produces a cocked-hat curve, then we can calculate different positions by using the standard deviation. Thus, if 100 animals were subjected to each dose, where the stimulus could just overcome average resistance, half the animals would be affected and the other half not. This would be the mode and, in a symmetrical distribution, also the mean. A dose 1 s.d. below this would only affect 16 per cent of the animals, and one 1 s.d. above the mean would leave only 16 per cent of animals unaffected. Similarly, a dose 2 s.d. below the 50 per cent level would exclude 1 in 40, or 2½ per cent, and a dose 2 s.d. above the 50 per cent level would leave all but 2½ per cent affected. It can be seen that if you plot the total number affected against the dose in logarithms it will produce a somewhat *S* shaped curve, which is spoken of as a sigmoid curve.

This correlation between dose and effect can be used in the other direction and where we have only 16 per cent of animals affected we can estimate that we are 1 s.d. below the mean. Thus by using the *t* table backwards, and halving the probabilities, we can correlate any percentage death with the distance in s.d.'s from the mean, and most of the deaths would lie between −2 s.d. and +2 s.d. To avoid these manipulations, and these negative figures, Fisher and Yates have published a table of Probits (Table IX) by which the correlation can be seen at a glance, and by adding the integer 5 to the 50 per cent level all the numbers have been kept positive. Using probits, 1 s.d. below the mean is probit 4, and 1 s.d. above the mean is probit 6.

To show the convenience of using probits let us take the simple example shown in the following table.

Correlation of Probits and Dose.

Dose	Shape of Cocked-hat Curve	Number of Animals Affected	Cumulative Percentage Affected	Dose as Logarithm	Cumulative Percentage as Probit
4	1	1	6·2	·602	3·5
8	4	5	31·2	·903	4·5
16	6	11	68·8	1·204	5·5
32	4	15	93·7	1·505	6·5
64	1	16	100·0	1·806	

It can be seen from this table that the dose goes up in a geometrical progression and the percentage animals that are affected goes up in the form of a cocked-hat curve, based on 4 rows of Pearson's wedges. This number of affected animals is cumulative because those susceptible at one dose are also susceptible to every higher dose. In the penultimate column we see the geometrical progression of the dose has been turned into an arithmetic progression by recording it in logarithms. And, in the final column, we see that the cocked-hat curve has been turned into an arithmetical progression by recording it in probits. Once both variables are recorded as arithmetical progressions they can be plotted as a straight line on squared paper, and the ordinary simple regression methods are applicable. Note that, because the dose that leaves all animals unaffected could lie anywhere under a certain level, and the dose that affects all animals could lie anywhere above a certain level, these doses are of very little help in fixing the position of the line, and therefore are usually omitted from the calculation.

Working example

To see how probits can be used we can use the results obtained with different doses of a bacterium, as recorded in Table 192. The table is based on the results of different experiments, on different days, and the number of live organisms in the dose—the viable count—was calculated after the injections had been given. This irregular scale of doses emphasises that it is not essential for the scale to be in a regular progression as long as it is expressed in logarithms, as they are in the second column. The number of mice inoculated at each dose also varies, but this irregularity is removed when each result is turned into a percentage, which itself is recorded as a probit in the penultimate column.

It is shown at the bottom of the table that the average probit is 5·34 and that the regression is about 1·7. If, therefore you wish to find the E.D.$_{50}$, that is, the Effective Dose that produces a result on 50 per cent of the animals, then from Table IX we see that we need probit 5·000. The simple regression formula of $Y = \bar{y} + b(x - \bar{x})$ will be translated as calculated death rate $=$ $5{\cdot}34 + 1{\cdot}7(x - 1{\cdot}96)$, so that to remove the 0·34 from 5·34 we require $1{\cdot}7\times0{\cdot}2$, giving $\text{E.D.}_{\cdot50} = 5{\cdot}34+1{\cdot}7(1{\cdot}76-1{\cdot}96)$. Therefore $x = 1{\cdot}76$, which is the logarithm for a dose of about 58 million organisms; note that we have coded by dropping the additional characteristic of 6 that would be required to record the doses in millions. From the error seen in the rest of the table it is obvious that a dose of 55 or 60 million organisms would be close enough to the calculated 58.

TABLE 192

Deaths due to the Intraperitoneal Inoculation of mice with varying numbers of Streptococci Strain 13, contained in 0·5 ml. of Broth Culture.

Viable Count 10^6 per ml.	Dose as Log (x)	No. of Mice Inoculated	Deaths Within 5 days	Per cent. Mortality	Mortality as Probit (y)	(xy)
5	0·70	50	1	2	2·9	2·030
25	1·40	75	29	39	4·7	6·580
75	1·88	59	37	63	5·3	9·964
150	2·18	63	53	84	6·0	13·080
250	2·40	70	57	82	5·9	14·160
350	2·54	57	51	90	6·3	16·002
450	2·65	50	45	90	6·3	16·695
Total	13·75				37·4	78·511
average	1·9643				5·343	11·216
(Items)²	29·9709					

x items squared $= 29{\cdot}9709$

$$\text{C.T.} = \left(\frac{T^2}{n}\right) = 27{\cdot}0091$$

True s. sq. for $x = 2{\cdot}9618$

Sum of products $= 78{\cdot}511$

$$\text{C.T.} = \frac{\Sigma x\times\Sigma y}{n} = 73{\cdot}466$$

True sum of products $= 5{\cdot}045$

$$\text{Regression} = \frac{5{\cdot}045}{2{\cdot}962} = \text{approx. } 1{\cdot}7.$$

LOGARITHMIC PROBABILITY PAPER

With the material recorded in Table 192 it can be seen that on two occasions the increase of 100 million live organisms failed to increase the death rate. Under these circumstances one gains little by using the most accurate method, and time is saved by using logarithmic probability paper.

An example of this paper is given in Fig. 193. One scale shows three cycles of logarithms and the other is marked out in percentages associated with the normal curve. For simplicity only a proportion of these lines are shown in the diagram.

The point given by the dose and the percentage deaths it effects can be plotted directly onto this paper. A free hand line can then be drawn to fit the points, as has been done in this figure. From this line it can be seen that an $E.D._{16}$ (or probit 4) would require about 14 million live organisms, $E.D._{50}$ (or probit 5) would need between 50 and 60 million, and an $E.D._{84}$ (probit 6) would require a dose of between 200 and 300 million.

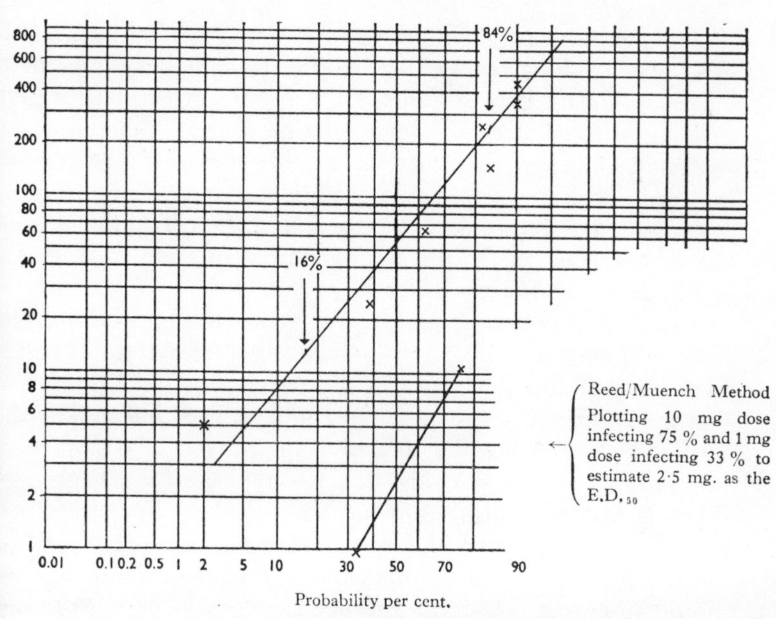

FIG. 193. Graphical method of finding percentage dose-effect.

METHOD OF REED AND MUENCH

When there are only a few doses in a geometrical series, and only a few animals at each dose, then an $E.D._{50}$ can be calculated by the simple method of Reed and Muench (1938). The limitations of the method is that it gives no standard error and does not lend itself to any more complicated technique for showing the difference between two different trials. As an example of the method we can make use of the data presented in Table 194

TABLE 194

Effect of varying doses of Mycobacterium Johnei *following intravenous injection into calves. (Dr. J. D. Rankin).*

Dose (mg)	No. of Calves Infected	No. of Calves not Infected	REED AND MUENCH METHOD		
			Total Infected	Total Uninfected	Percentage Infection
100	2	2	8	2	80
10	4	0	6	2	75
1	2	2	2	4	33
0·1	0	4	0	8	0

This method is based on the ingenious idea that if any individual dose would infect, say, two calves, then any higher dose would also have infected them. Similarly, if any calf survived a high dose, they would also have survived a lower dose. The idea that one animal at a high dose may accidentally survive or one at a low dose might accidentally die, is balanced as much as possible by making the number of doses on one side of the $E.D._{50}$ the same as the other, although this may entail throwing away some of the results at one end or the other.

Applying the method, we see that at the 0·1 mg. dose at the bottom of the second column, no calves were infected; 2 calves were infected at 1 mg. and 4 at 10 mg. Using our assumption that the 2 infected at 1 mg. would also have been infected by the 10 mg. dose, the figure for 10 mg. is raised to 6. Two more were infected at 100 mg. but this dose would also have infected the other 6 so the total at 100 mg. is 8. The cumulative total showing the number of animals remaining unaffected is worked out in the same way beginning at the top of the column. From these two columns of totals the percentages can be worked out and this is presented in the last column, thus, in the first line, 8 infected and 2 not infected gives an 80 per cent infection rate.

From the percentage column it can be seen that the $E.D._{50}$ dose must lie somewhere between the 10 mg. dose that infects 75 per cent, and the 1 mg. dose that infects 33 per cent. If we plot these two points on our logarithmic probability paper (see Fig. 193) we can see that the $E.D._{50}$ will be about 2·5 mg. of dried culture, but in the Reed Muench method advantage is taken of the fact that in the probability paper the distances between 20, 30 and 40, up to about 80 per cent are roughly equal, and therefore the percentage is roughly proportional to the distance. If the percentage is proportional to the distance, and if the distance between 10 mg. and 1 mg. is 75-33 per cent, then to obtain the E.D.$_{50}$ we will want the proportional distance of:

$$\frac{50-33}{75-33}, \text{ or } \frac{17}{42}, \text{ or } 0{\cdot}4$$

To get a straight line we have seen that our dose usually requires to be stated as a logarithm. If we have logarithm paper we can easily measure off 0·4 of the distance between the two doses by using a ruler. Alternatively, if we wish to do it by arithmetic, we can turn the doses into logarithms so that 10 mg. equals 1·0 and 1 mg. equals 0·0. The difference is obviously 1·0, and 0·4 of this is 0·4. The E.D.$_{50}$ will therefore be the antilogarithm of 0·4, that is 2·5 mg. Note that the statement that: " The E.D.$_{50}$, by the method of Reed and Muench, was 2·5 mg.", might sound like a sacred truth to the uninitiated, but the table shows an actual E.D.$_{50}$ at both 100 mg. and at 1 mg., and all that has been done is to make the arithmetical best out of scanty data. In point of fact, subsequent work showed that this estimation was very accurate in this instance.

Collection of animals and materials

Having dealt with measurements we can now turn to a general consideration of preliminaries. Notice must be given of the number of animals that will be required. If these animals are not to be supplied by the laboratory then they must be bought sufficiently early to permit them to settle down to a new environment before the experiment begins. This period will also permit one to see that they are free from lethal infectious diseases, or to discard any that are clinically unwell.

Chemicals used for tests should all be of the same batch issued by the manufacturer, and, if not, the different batches should be mixed. This ensures that no additional variable is added during the course of the experiment, and makes one correction term cover the whole bulk. Similarly, when using living organisms for test doses, sufficient tubes of the same culture of organism should be dried off, and sufficient quantity of the same brew of bacterial media should be stored in the cold room, to try to standardise the factors required to produce the same level of pathogenicity.

Instruments and techniques should be examined for their reliability and, if possible, really important measurements should be estimated by two different methods. An example of this would be to measure a substance in blood serum by a photo-electric method and in a colorimeter, so that neither an increase in turbidity nor an increase in the colour of the serum would be blindly accepted as an increase in the substance to be estimated. Further, should the more elaborate instrument break down, you will still obtain some answer to what might be a costly experiment.

Time of year

The time of year may play an important part in the success of an experiment. Thus several years ago a feeding stuff merchant was sued because several animals died after being fed on his produce. He called in an "expert witness," who prepared evidence by feeding other animals on fodder containing the same plants cut at a time of the year when he knew they would be harmless, and was able to fool the judge that the original produce was harmless. This is the complete prostitution of science, but it does emphasise that the time of year can be very important. Less spectacular is the fact that in some environments herbivorous animals will show a slight anaemia, or fail to recover from a severe one, during the winter months, while guinea pigs may suffer death from endemic diseases, hidden through the rest of the year, during the months in which their diet is low in vitamin C.

The normal breeding seasons of different animals must also be taken into consideration and, where breeding is to be encouraged outside the breeding season, help must be given by injecting gonadotrophic hormones, or by providing extra warmth and light to laboratory animals.

A general discussion on the selection of materials and the importance of finding a suitable laboratory animal, both important points when digging the foundation, was given in Chapter 8.

Choice of test dose

As we dealt with the estimation of test doses earlier in this chapter it is convenient to deal with the choice of a test dose here, rather than in the next chapter, where it more properly belongs. Where a test dose is required, the obvious choice is an $E.D._{50}$ (Effective Dose affecting 50 per cent of animals) or an $L.D._{50}$ (Lethal Dose killing 50 per cent) for this is the most sensitive dose because it does not rely on the chance presence of exceptionally resistant, or susceptible, animals. When trying to show a slight change from a previous estimation of toxicity, the $E.D._{50}$ is therefore the dose to choose.

There is, however, the possibility that you are not interested in small changes, and might wish to test a substance to see if it has a definite protective power, using 8 animals in each group. Under these circumstances the theoretical value for the control group when using an $E.D._{50}$ would be 4 affected and 4 not affected. The best result for the treated group would be none affected, or 0:8. Mainland's tables show that 4:4 compared with 0:8 has a probability of $p = 0{\cdot}077$ and is not significant. If an $E.D._{84}$ had been used the theoretical expectation would be 6 affected and 2 not affected, compared with none affected, provided the treated group was still protected at this higher test dose. This would give 6:2 compared with 0:8 giving a probability of 0·007, a result that is more than 10 times more significant than the first one. In a similar way, if the factor to be tested made the animals more susceptible, then an $E.D._{16}$ might be indicated.

In other circumstances where the material was known to be very variable, or the test dose unreliable, then it might be worth dividing the animals into 3 sub-treatments and using the 3 test doses, namely $E.D._{16}$, $E.D._{50}$ and $E.D._{84}$. These 3 doses could also be used when one wished to see the regression of the result with the increase in dose. Thus although it can be said that in general test doses are scaled to an $E.D._{50}$ there is no reason why higher, or lower, doses should not be used in particular circumstances, and no rule to prevent you using an $E.D._{30}$ or an $E.D._{75}$ if you have some reason for it.

It was said that the cumulative percentage of animals affected usually gave what was called a sigmoid curve when the dose was increased as a geometrical progression, but with some lethal exotoxins this percentage rises so steeply that it is almost a straight line, and here it is impractical to try to find an $E.D._{50}$. Where this is the position, use is made of a minimum lethal dose (m.l.d.), which is the smallest dose that can be relied upon to cause the death of all the animals in the test. Using this dose the survival of even 1 animal suggests the presence of some added immunity, and the statistical significance of its survival will depend on the large number of animals that were used in estimating the m.l.d.

FURTHER READING

Instructions in this chapter have been limited to finding a suitable test dose, but specialists sometimes require to compare one line with another, to judge if they represent samples from the same population. This entails finding the standard error and involves a complication in that samples at the extremes are less reliable than those near the mode, because the extremes depend on the presence of exceptional animals.

Miller and Tainter (1944) give a graphical method, using semi-logarithm paper and recording the percentages arithmetically as probits. They indicate the approximate reliability of different probits by stating that, where the number of animals at each dose is equal, doses falling at probits 4 and 6 have only two-thirds the reliability of that at 5, and those at 3 and 7 only one-fifth of the reliability.

The simplest method is to use logarithmic probability paper and follow the instructions of Litchfield and Wilcoxon (1949) who have done most of the work for you. Alternatively, if the experiment concerns the numbers of animals dying over a period of time, which in some diseases also produces a sigmoid curve, then there is a similar method given by Litchfield (1949).

The most accurate method depends on the use of logarithms and probits, with the different percentages weighted according to their reliability, and this is explained in the second edition of Mather, but it must be emphasised that reliability is dependent on the use of a large number of animals, and therefore is of most use when dealing with such animals as insects. For specialist work, involving covariance and factorials, there is a book by Finney (1947).

SUMMARY

Before starting an experiment one should have some idea of the error in the estimations, checking systematic error by using another method and random error by replicates on the same sample, or duplicates on a number of samples. Error can be cut down by always using the same apparatus; always using the same batch of material, or using different batches that have been well mixed together; and by having the results read by the same person. Random error can be cut down by taking care and by replicate estimations on each sample or, preferably, taking many samples, so that the results can be averaged, working on the basis that the error is halved where 4 times the number of samples is estimated. An important experiment should not depend on a single instrument.

An E.D.$_{50}$ can easily be found by using logarithmic probability paper, and where small changes are being looked for it is the most sensitive dose to use. With more definite changes, an E.D.$_{16}$ or an E.D.$_{84}$, according to circumstances, may be useful.

References

FINNEY, D. J. (1947). " Probit Analysis " Cambridge Univ. Press.

LITCHFIELD, J. T. (1949). " A Method for Rapid Graphic Solution of Time-per Cent. Effect Curves." *J. Pharmacol. exp. Ther.*, **97**, 399.

LITCHFIELD, J. T., and WILCOXON, F. (1949). " A Simplified Method of Evaluating Dose-effect Experiments." *J. Pharmacol. exp. Ther.*, **96**, 99.

MILLER, L. C., and TAINTER, M. L. (1944). " Estimation of the ED$_{50}$ and Its Error by Means of Logarithmic-Probit Graph Paper." *Proc. Soc. exp. Biol.* (N.Y.), **57**, 261.

REED, L. J. and MUENCH, H. (1938) *Amer. J. Hyg.* **27**, 493.

CHAPTER 11

CHOICE AND ALLOCATION OF ANIMALS

All our physicians cannot tell what an ague is, and all our arithmetic is not able to number the days of a man; which, God knows, is not the fault of arithmetic.

SAMUEL PEPYS

In the last chapter we dealt with preliminary measurements made on materials, or apparatus, and we can now turn to the allocation of animals into the different groups.

NUMBER OF CONTROL ANIMALS

The number of animals used as controls depends very much on the knowledge that is available on the materials that are being used. For example, if you are using a substance for which the m.l.d. (minimum lethal dose) or m.e.d. (minimum effective dose) is known, then, using this dose for the type of animal on which it was based, you will know that every animal will be killed, or affected, unless the treatment has in some way lessened its toxicity. When working with an m.e.d., therefore, all that may be necessary is the use of one control animal just to show that at least the substance came from the correct bottle, or the use of two or three animals when the result will represent an important logical step in building up an argument; to use a larger number of animals would merely show a lack of compassion.

In contrast, where the conditions are such that the test dose, or the treatment, is not always obviously effective, then it is necessary to try to judge the result by using a test of significance on the data, which may be the enumeration data of the "pass," "fail" type, or may consist of measurements.

The most efficient experiment where the results are of the "pass," "fail" type is one that shows the disparity between the two groups most clearly by producing the highest χ^2 value. It can be seen from Table 201 that the highest χ^2 value is obtained when the control and experimental groups are equal in number. This table is based on the assumption that there are 64 animals available, that one group will show 50 per cent of animals affected and the other group only 25 per cent. Inspection of the table shows that the highest χ^2 value exists where there are

TABLE 201

Effect of varying the number of animals allocated to the experimental and control groups. Where the total available is 64, the death rate in the control group is 50 per cent. and in the treated group is 25 per cent.

RESULTS

	Control Group		Treated Group			
Number of Controls	Dead (50%)	Alive (50%)	Dead (25%)	Alive (75%)	χ^2	p
8	4	4	14	42	2·2	0·138
16	8	8	12	36	3·5	0·061
24	12	12	10	30	4·2	0·040
32	16	16	8	24	4·3	0·038
40	20	20	6	18	3·9	0·048
48	24	24	4	12	3·0	0·083

Note that χ^2 reaches its highest value when control and treated groups are equal. Yates' modification was not used.

32 controls and 32 experimental animals, and that there is a fall in value as the inequality between the groups gets larger. In general, therefore, one aims at an equal number of animals in the control and experimental groups, and not only does this give the highest significance where there is a disparity, but it allows a direct visual comparison without mental gymnastics.

In experimental work, however, we are not always looking for the highest significance, we may be looking for the best value for our money. To illustrate this let us take the warble fly that lays its eggs on the hairs on cows' legs. The larvae from the eggs penetrate into the skin and enter the body. After growth each larva finally gains freedom by bursting through the skin of the back, destroying the value of the hide. Let us assume that we wish to test a substance that repels flies when sprayed on the cow's skin, and that there are 100 cows at pasture and available for the experiment. Assume also that from previous experience we judge that about 8 out of 10 cows will be affected if we do not protect them, and that a repellent will not have much practical value unless it protects 6 out of 10 cows. Under these conditions, if we divide the cows into 50 controls and 50 experimental animals, and if the results give the ratios mentioned above, we would have the following four-fold table:

Cows	Affected	Not affected	Sub-totals
Controls	40 (80%)	10 (20%)	50
Experimental	20 (40%)	30 (60%)	50
	60	40	100

$\chi^2 = 15{\cdot}04$. From graph given by Yule and Kendal $p =$ apprx. ·0001.

Routine research investigation is a step by step process, with each step tested for significance before its inference is accepted. If the above test was just one step in a series then, by splitting the cows into two equal groups we have obtained a probability of 1 in 10,000, which is extravagantly low, and quite unnecessary. Under these conditions there would be two alternatives. Firstly, if in its present stage of development there was a danger that the repellant was toxic to the cow, or if it were very expensive, or tedious to apply, then it would be a business proposition to test it on only 20 cows, leaving the 80 to give the stability of a large control group. This alternative would give a highly significant result with an effective repellent even if there was somewhat less disparity between the groups than was anticipated, and although only 20 cows were put at risk. Secondly, if the repellent was known to be non-toxic, was cheap, and was easy to apply, then it would be a business proposition to leave only 20 cows untreated. This would also provide a highly significant result with an effective repellent and would also hold out the hope that over half the hides would be protected.

This test, however, might not be part of a step by step process, but might be the final check before issuing the repellent to the public. Here the circumstances are quite different and we would again place half the cows in each group and use as many cows as possible, but here the object is not to obtain an extremely high degree of significance but to show that about the same number of unexplained diseases and deaths occur in each group. For this purpose, in which the event to be measured is likely to be rare, the half and half design is the most sensitive and the most logical.

The same principles apply to experiments that involve measurements rather than events, and in any case of doubt equal numbers should be used in each group, if only because we are practising a crude " method of difference " argument, and what better logical control can there be for 7 fat cows and 7 lean ones than another 7 fat ones and another 7 lean ones? Finally, when dealing with a factorial design involving measurements, then for the mathematical interpretation by an analysis of variance every group must contain the same number of animals as there are in the other groups opposed to it.

Selection of animals

There is a story told in this country (in France it is probably told differently) of an Englishman in France who is taken for a little rough shooting. While walking through a wood a nice plump rabbit jumps up and scurries away, the Englishman raises his gun but the Frenchman cries out " Don't shoot! Don't shoot! That is Jean, we never shoot Jean! " A little farther on, out runs another young rabbit, and the Frenchman shouts, " Don't shoot! Don't shoot! That is Marie, we never shoot Marie! " A little later as they walk on an old rabbit with one ear missing and scars all over his body, hobbles out on three legs from behind a tree. The Englishman just looks at it but the Frenchman cries out " Shoot! Shoot! We always shoot Pierre! " A similar situation often arises when you join the staff of a laboratory and you are taken to see the experimental farm. As a research worker you may imagine that these animals are for use, but when you have planned your first experiment you will find that these animals are inviolable, and are kept for the delectation of patrons and visitors, but that, somewhere there is an old shed where " Pierre " is at your service. The question of " Pierre " is a big one, for it would seem that all over the world workers are paid to work on animals without being provided with the animals to work on.

Thus it can arise that after great effort you obtain 9 animals that appear reasonably healthy and 1 that appears sick. What are you to do? To put " Pierre " into the group that you expect to deteriorate would be to deliberately bias your result in the direction you predict. To put him, as an act of faith, into the other group is ethically less reprehensible, but you must realise the risks you run. If this animal dies then your result may read " Four animals improved and one died." If the experiment involved a measurement made on the live animal then no matter how you juggle with figures, there is no ethical way of producing such a figure for a dead animal; you must either conceal the fact that you used this animal, or indulge in a lengthy explanation as to why you used this animal under the proviso that you were going to accept the result if it was what you wanted it to be, and reject it if it went against you. Distribution to one group or the other by chance is no way out, for the method does not read, " Distribute five suitable animals in one group and four suitable animals and one unsuitable animal into the other group, by a method of chance."

In drawing conclusions from an experiment, 1 elephant is equal to 1 mouse, and if you require 30 mice before you can anticipate a significant result then you will require 30 elephants before you obtain the same significance under similar circumstances. This emphasises the need, in preliminary work, of finding an inexpensive animal to deal with, for the idea that you should get 100 times more information from a large animal, because it is 100 times more expensive, dies slowly in the minds of those who must provide the money. But, in spite of cost, "Pierre" must be refused admission, for it is better to work with 1 animal missing, rather than to be confused by the queer information it might provide.

It is obvious that this rejection of unsuitable animals should take place before the experiment starts but there is a true story of a post-graduate student who told his professor that although some of his treated animals had become sick so, also, had some of his controls. The professor was rather indignant about this. "You must not use sick animals as controls!" he said, and promptly picked up the sick ones and killed them. This happened several years ago, but because it happened at all it is worth repeating that the time to reject sick animals is before the experiment begins, but once the experiment has begun then, in army slang, " Once they're in, they're in! "

Bias in animals

It is sometimes said that soldiers, left on their own initiative, can react to upright objects in only three ways, they can lean against them, kiss them, or salute them. The experimenter, similarly, has three alternative actions for dealing with sources of bias, likewise suggesting contempt, interest or respect.

As an example of the first alternative presume that the animals available for an experiment consist of more than one breed, for example, albino, brindle and black strains of mice; similarly they might be of one breed but of different ages, such as young, adult and aged. In these circumstances, providing that you decide that these differences will not produce an obvious difference in reaction, the simplest method of dealing with the possible bias is to allocate the mice, regardless of breed, into the different treatments by using random numbers. By this method both the possible bias in breeds and the unknown biases in the individual mice are distributed by chance, one bias augmenting or modifying another. Thus, although the sum of squares between treatments might be increased if a breed bias is present, the statistical analysis is valid.

The second alternative occurs where it is realised that different breeds might show obvious differences in their reaction to disease or other experimental interference. Here one can measure the difference between breeds by making each breed into a sub-class, for by allocating the same proportion of, say, black to albino breeds to each treatment it will still be possible to add and subtract the sums of squares in an analysis.

Thus one might have 3 treatments, A, B and C, with 8 albino, 5 brindle and 3 black mice in each.

It is possible to take advantage of this restraint even although the number of mice in the different breeds are not the same. Thus, as we did with the randomised blocks of milk yields in Chapter 8, we could take away the s.sq. between the different breeds and thus decrease the s.sq. for error. In addition, although the fact that the number of mice was not the same in each cell, which destroys the true factorial design, we have the same proportion of each breed in each treatment, and this permits us to test the interaction between breed and treatment, and to discover, say, that one breed is very susceptible to a certain treatment. This type of analysis comes under the term Proportional Sub-class Numbers (see for example, Snedecor) and the method is also applicable to experiments where one had, say, 16 albino, 10 brindle and 6 black in one treatment and 8 albino, 5 brindle and

3 black in another, for the criterion is that sub-classes are proportional in each treatment, and not necessarily the same size.

The third alternative is where it is necessary to use animals of different breeds although their numbers do not permit equal proportions for each treatment. In these circumstances, if we wish to treat the bias that may be concealed in different breeds with respect, we do not mix the mice of the different breeds together and then randomise them to the different treatments, as we did in the first alternative, but we randomise the mice from each breed separately. Thus, if one had 4 treatments and 17 albino, 9 brindle and 6 black mice were available, then we might begin with the albino mice. Dividing random numbers by 4 the mice would be distributed to the 4 treatments according to whether the differences left over by the division was 1, 2, 3 or 0. With 4 mice in each treatment the remaining mouse could be allocated by chance to any of the 4 groups.

Starting again the brindle mice would be allocated 2 to each group and the extra mouse allocated by chance to 1 of the groups containing 6 mice. Finally, 1 black mouse would be allocated to each treatment and the remaining pair allocated at random into the 2 deficient groups.

As each mouse would have an equal chance of falling into any group the items are legitimately randomised yet the bias has been spread as evenly as possible. If, however, after the experiment was over it was found that breed differences could not be ignored then it would be a poor design to analyse, and would require the technique for Disproportionate Sub-class Numbers, in which the harmonic averages of each breed in each treatment are used to apportion breed differences but the value for every mouse is used to find the total sum of squares. This method of apportioning breeds as equally as possible to each treatment is sometimes spoken of as " purposive selection " (Mainland, 1952).

By applying these restrictions the amount of variation between the average of one group and that of another is cut down and, more important, the suspected bias is decreased. On the other hand if there are several groups and the results will be tested by the analysis of variance, then it must be remembered that the total sum of squares is fixed by the result and the less the sum of squares between the averages for the different treatments the larger is the sum of squares for error. This would seldom be a serious disadvantage and most experimenters would prefer the logic of controlling the bias to a possible small decrease in the sum of squares for error.

Applying more than one restriction

We have seen that where we use material that may contain a bias, we can, (i) randomise it, (ii) make it into a sub-class of its own, or (iii) restrict randomisation. The last action is roughly a compromise between the other two and, like most compromises, possesses practical utility. When, therefore, the interest is in the practical effect of a treatment, and there are several differences in the attributes of the animals to which it will apply, it is usual to include these different types of animals but to apply as many restrictions as are practical.

This process can be shown more clearly by following an example than by stating rules. Let us therefore assume that we are going to set out an experiment to test treatments that might increase the growth rate of young pigs, about 10 weeks old. In setting out the experiment we must take into account that some litters will naturally grow faster than others, that those individual pigs that have already attained a high weight will probably continue to grow quickly, and that castrated males grow faster than the females.

We could easily select male and female treated groups and compare them with male and female control groups, but it would be much more difficult to deal with the different litters and the different weights in this way. In a test of this type we must decide, not only what sources of bias there are, but the priority that we should give to them. In this instance both litter and weight appear from past experience to be more important than sex, so that it would be logical to restrict the litters and weights and hope to distribute the different sexes more or less equally in the randomisation.

Let us assume that we intend dealing with the following litters, which are recorded with their weights in an array:

Litter 1	wt.	33	39	40	40	42	44	49	50	52	52	
	sex	m	m	f	f	f	m	m	m	f	f	
Litter 2	wt.	39	41	43	44	48	50	51	52			
	sex	m	m	f	f	m	m	f	f			
Litter 3	wt.	37	39	40	42	43	43	44	45	50	51	52
	sex	m	m	f	m	m	f	f	m	f	f	f

Suppose that we wish to divide these pigs into three groups. The greater the similarity in the weights of the groups, the more sensitive the comparisons between them. We might therefore, start by discarding the pig at 33 lb. in litter 1 and the one at 37 lb. in litter 3. We could then start splitting the remaining weights into lots of 3, so that we could randomise one from each lot into each of the main groups. We find that we cannot do this perfectly because there are 8, and not 9, in litter 2; to overcome this deficiency we might take 1 from litter 3. This might give the following table:

Litter 1			Litter 2			Litter 3		
39	42	50	39	44	50	39	43	50
40	44	52	41	48	51	40	44	51
40	49	52	43	(43)	52	42	45	52

The transference of one pig of 43 lb. from litter 3 to litter 2 is indicated by brackets. We have now got the items stratified into litters, with each litter stratified into low, medium and high weights. The next task is to randomise each column of three items into the three groups. One way of doing this would be to work along the top row and, by using a dice, shuffled cards or random numbers, allocate the top items into one of the three groups. With this done it would be only a question of tossing a coin, or reading off odd or even random numbers, to allocate the second row to the left or right remaining groups. The third row would then fill the empty places. By such methods the items were allocated as follows:

Litter	Wt.	Group 1	Group 2	Group 3
1	L	40F	39M	40F
2	L	43F	39M	41M
3	L	42M	40F	39M
1	M	49M	44M	42F
2	M	44F	(43M)	48M
3	M	45M	43F	44F
1	H	52F	52F	50M
2	H	51F	50M	52F
3	H	50F	51F	52F
Totals		416	401	408
Average		46·2	44·6	45·3
After manipulation. (Interchange of 43F and 39M).				
Totals		412	405	408
Average		45·7	45·0	45·3

We have now divided the pigs into three groups. By using purposive selection each group contains, as far as was possible, an equal number of animals from each litter, and an equal number of pigs at low, medium and high weights. Yet the final allocation of any individual animal to any particular group depended on chance alone.

We have used restrictions on both litters and weights and have left the distribution of the sexes to randomisation. This is a satisfactory design and we can now carry out what must always be the last randomisation, and assign the different treatments to the different groups.

INTERCHANGING

The above design, although orthodox from the mathematical point of view, might be queried from the logical point of view. Thus the experimenter might argue that as sex represents a bias he cannot purposely leave 3 males in one group, 4 in another, and 5 in the third, when by a single interchange he could have the logically irreproachable comparison of 4 males and 5 females in each group. Further, if his interchange was with the underlined figures, 43F and 39M, he could have not only 4 males in each group but also 1 of the small 39 lb. pigs, and bring the average of each group to about 45 lb.

This interchange to make the groups more equable is not acceptable to all statisticians and some would argue that the whole basis of a statistical test was that the items had been distributed to the different groups by chance alone. And that to distribute them by chance and then manipulate them into something different was to make the whole structure invalid. It does appear, however, that two or three interchanges, depending on the size of the experiment, are becoming acceptable. If one does not permit such interchanges and by using random numbers one fills one group entirely with male animals, would this be logically satisfying? Unfortunately the books on applied statistics with which I am familiar avoid this question. Mainland (1952) states that: " When an investigator employs a method that is not strictly random as if it were equivalent to a random technique, the onus is on him to prove it justifiable, by experiment, not argument: and this, even if possible, would entail a very large investigation." As against this it could be said that if you interchanged items to try to remove a bias, and by doing so moved the unknown factors attached to each item then, because these factors were quite unknown to you, their ultimate position was, to all intents and purposes, dependent on chance. Until the more elementary books give instruction on this point, it is a problem on which to ask assistance from a professional biological statistician, before the experiment is begun!

Using random numbers

In order to save the time that would be spent by tossing a coin, throwing a dice, or shuffling numbered cards, some books on applied statistics provide a set of random numbers. Those given by Fisher and Yates in Table XXXIII show a random arrangement of numbers from 1 to 100 in blocks of 25. To illustrate their use let us suppose that we are going to randomise 7 litters of 8 pigs into 8 different groups. One way to do this would be to start with one litter and select some point among the random numbers from which we could work downward, upward or sideways. If the first random number was 53 we would divide this by 8 leaving a difference of 5, and we would place that pig into group 5. If the next number was 23 the second pig would go into group 7. After a time we would get a further difference of 5 or 7 but as one pig had already been allocated to these groups we would ignore these numbers and try the next one; this would involve a waste of time and mental arithmetic. When we had allocated all the pigs in one litter we would continue in the same way with the others until all the pigs had been assigned. In doing this we would bear in mind that as the random numbers ran up to 100 (00) and as 12×8 were 96, we would ignore 97, 98, 99 and 00 to avoid a bias towards groups 1, 2, 3, and 4.

You can avoid the waste of time caused by getting the same differences more than once if you adopt a slightly different method. Arrange the 7 litters in columns side by side, with the 8 pigs in each giving the 8 rows. Begin by going from litter to litter along the top row allocating a pig from each litter to one of the 8 groups by dividing the random number by 8. It does not matter if more than one pig falls into a group because they will be of different litters and each litter will be kept on a different line. Because only 1 pig from each litter must enter each group there will now be only 7 vacancies to be filled, so that mentally renumbering the groups from the top down we can divide the random numbers by 7 to allocate the second row. This will leave only 6 vacant places so that we can continue diminishing the divisor of the random numbers until the final row, which will be allocated to the only vacant places. Note that you ignore figures over 96 when dealing with 8 and 6, and 99 and 00 when dealing with 7, and so on. This illustration should suggest other methods of employing the same principle.

Number of animals to be used

There are many practical points to be taken into consideration when deciding on the number of animals to be used, and sometimes the greatest wisdom is shown by refusing to start an experiment because the number of animals available are insufficient to secure a reliable result.

Some idea of the number of animals can often be obtained by reading up previous experiments on the same or a similar subject, but where the experiment is easy to arrange and the conditions are stable, then the easiest method is to carry out a small pilot experiment with only five animals in each group, in order to form some opinion of the strength of the experimental interference and to have a basis for calculating the numbers that might be necessary for the main experiment. The planning of a pilot experiment that cannot give a significant result is quite laudable, except where the result is spoken of with pride without any effort being made to carry out the main experiment to try to verify it.

In contrast, where the experiment depends on some variable condition such as an epidemic of disease, then you must work out the best possible experiment using whatever information you have, and then put in all the animals that are available in the hope that you will get a highly significant result, for there may be no opportunity of ever repeating it.

It can be seen that so much depends on practical conditions that very often there is no opportunity to calculate what is a desirable number of animals for a group, but when theory can be used it has a practical objective and can prevent you wasting your time, and other people's, either by planning an experiment that has very little chance of providing a reliable result, or by planning one that would be wasteful of material.

The technique for predicting the number of animals that are required is different for enumeration data (Present or Absent) and for mensuration data (How much is present) and the two types will be dealt with separately.

Numbers in enumeration experiments

Help in deciding the number of animals in an experiment of the qualitative type can be had either by using the χ^2 test, by looking possible results up in Mainland's tables (Mainland 1948 and 1956), or by using binomial paper, and some indication will be given as to how these different methods may be used.

Chi-squared Method. If we look back at Table 201 it can be seen that if there were 50 per cent affected in one group and 25 per cent affected in the other, we would need 64 animals to obtain a minimum of significance, with $p = 0{\cdot}038$. This would permit us to go on to the next experiment but would certainly not be reliable enough to justify advocating the treatment to the public. Thus to show a highly significant difference between proportions of 50 per cent and 25 per cent requires over 60 animals. Suppose that, before advocating our treatment to other people, we would like to have a highly significant result, that is $p = 0{\cdot}01$, how many animals would we need for this?

Mainland (1938) points out that where Yates' modification is not used, the χ^2 value is doubled if you double the value of the figures in each of the 4 cells. If we act on the assumption that in the next experiment the proportions will be just the same, then we can say that as our 64 animals gave a result of $\chi^2 = 4{\cdot}3$, then with twice this, the result for 128 animals will be $\chi^2 = 8{\cdot}6$. If we wish to use the first experiment to predict the result of the second, for which we want a result of $p = 0{\cdot}01$, we look up the χ^2 value for $p = 0{\cdot}01$ and find the figure $6{\cdot}635$. This figure is about $1{\cdot}54$ times $4{\cdot}3$, so that we would want $1{\cdot}54 \times 64$ animals, or about 100 animals, divided into two equal groups, in order to have a reasonable chance of getting a result near to $p = 0{\cdot}01$.

Binomial Tables. The second method for judging a suitable number of animals is to use the tables introduced by Mainland (1948 or 1956). These tables give the one-sided probability for four-cell tables, and to get the double-sided probability in which significance has the conventional limit of $p = 0{\cdot}05$ the tabulated results must be doubled. Some abridged results for a part of the table are given below, corrected for double-sided probability.

TABLE 213

Selected results of exact tests on four-cell tables taken from Mainland's tables.

Number in each group	Group 1 A.s	Group 1 B.s	Group 2 A.s	Group 2 B.s	Double-sided probability
4	0	4	4	0	0·03
5	0	5	5	0	0·01
	0	5	4	1	0·05
10	0	10	5	5	0·03
	2	8	8	2	0·02
20	0	20	6	14	0·02
	6	14	14	6	0·03

From Mainland's tables it can be seen that with 3 animals in each group it is impossible to get a significant result, for if all 3 in one group are A.s and in the other group all B.s this could occur one in ten times by chance alone. Remember, however, that there is no law that you must only take guidance from a significant result; the statistical test gives you the odds and offers you $p = 0{\cdot}05$ as a conventional guide. If you feel that odds of 1 in 10 is sufficient encouragement, then introduce the next modification and go ahead.

Looking at Table 213 it can be seen that with 8 animals, 4 in each group, all in one group must be A.s and all in the other group B.s, if the result is to be significant. Five in each group is the first total in which even one animal can be misplaced and the result remain significant. With 10 in each group an interesting criterion to bear in mind is that the result 2 and 8 compared to 8 and 2, is significant. With 20 in each group you require 30 per cent A.s in one group and 70 per cent A.s in the other group to show a difference.

By looking at these tables and estimating the sort of result you expect to get, you can get some idea of the number of animals you will need; unfortunately this is often twice the number you hoped to get away with.

A second type of problem concerns the comparison of a foreseeable result with some theoretical percentage. For example, if you have obtained 20 ewe lambs and only 12 ram lambs by a certain method, how many more must you obtain by this method before you can accept that you have upset a 50 per cent ram and ewe ratio? This type of question will be considered when dealing with binomial paper, and it is only necessary here to point out that Mainland does provide tables that are equivalent, and more accurate, than measuring from the " split " in the binomial paper.

Binomial Paper. The idea of this paper was introduced by Fisher, and the modification and use of this paper is described in a lucid article by Mosteller and Tukey (1949) and, if possible, this article should be consulted. The paper is constructed on the basis that both vertical and horizontal scales are measured in centimetres, and these distances are recorded as the squares of the numbers they measure (see Fig. 215). The points marked 100 are joined by a quadrant, of which one use is to measure percentages, thus for 40 per cent., the 40 on the vertical scale is followed until it touches the quadrant, and from here dropping vertically down the horizontal scale will read 60. By drawing a line from 0 through the 40 per cent point on the quadrant, a line is produced in which all the points on it are in proportion of 4 on one scale to 6 on the other scale. Mosteller and Tukey refer to this as a 40 per cent split.

The paper is used for problems that depend on the binomial distribution and for which we have used the χ^2 test and the exact test, and for which some people use a formula based on a standard deviation of $\sqrt{npq}$. The significance of the result is measured graphically by a scale of standard deviations given at the top of the sheet. Two s.d.s measure about 1 cm. in length, and in the article already mentioned a table gives the values for proportions of a centimetre.

In addition to the line giving the split, use is made of actual results, such as 37 affected and 10 non-affected, and these are plotted as paired counts. In a similar manner to the use of Yates' modification in the χ^2 test, each of these points is plotted, not only as the number itself but as the next higher number. Thus the paired count 37, 10, is marked by placing a dot at 37, 10, and also dots at 38, 10, and 37, 11. These dots at low values give a tiny triangle, and Mosteller and Tukey suggest that for general work the mid-point of the hypotenuse is a useful point from which to measure significance.

To see how the paper is used for prediction, let us assume that we are working with a factor which we think is inherited in a simple Mendelian manner. Investigating the progeny we find that the factor is present in 20 and absent in only 12. We wish to know how many more progeny must be born before we can say that our sample comes from a population in which A.s represent either a half, or a quarter, of the total.

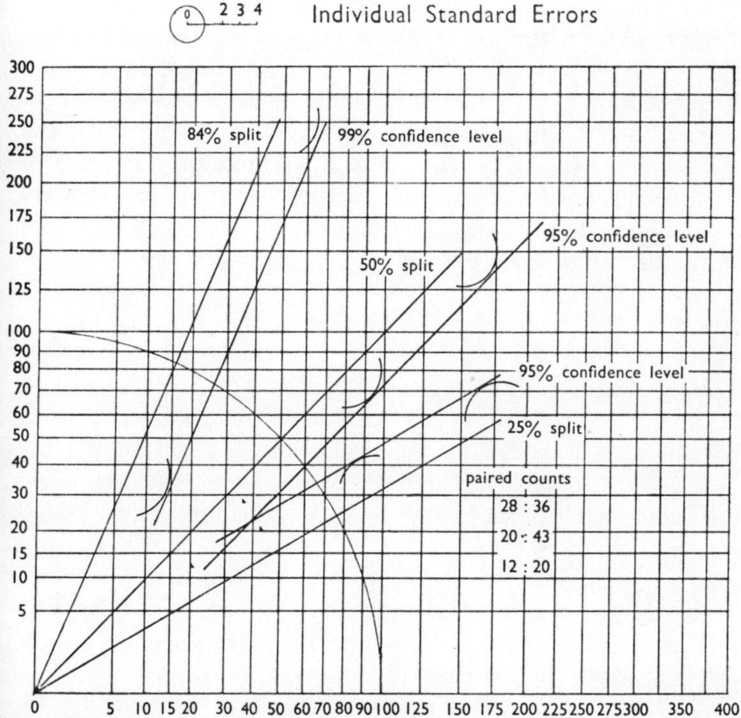

FIG. 215. Use of binomial probability paper to estimate the number of experimental animals required.

To estimate this, we draw the 25 per cent split and the 50 per cent split, as shown in Fig. 215. Parallel to these lines and on their inner side, we draw the 95 per cent confidence levels at 2 s.d. (1 cm.) from the split. If we wish we can now plot our paired count of (12, 20) (13, 20) (12, 21) to see where it lies, and it can be seen that it lies within the 95 per cent confidence level of both the quarter split and the half split, confirming that at the moment there is too little evidence for a verdict to be given.

In the original of Fig. 215 the 95 per cent confidence lines appeared to meet at the point given by the paired count of (23, 39), with the triangle (22, 40) (22, 39) (23, 39) below and the triangle (23, 39) (24, 39) (23, 40) above. From this we can say that 63 animals (either 23+40 or 24+39) would, with a 95 per cent confidence level, be sufficient to show if the factor was inherited by half, or by only a quarter of the progeny.

To test this with a few trials, it will be seen that 20 and 43 is homogeneous with the 25 per cent split but significantly not 50 per cent, and 28 and 36 are homogeneous with the 50 per cent split and significantly not 25 per cent. Accepting our prediction, and accepting the fact that we already have 32 progeny, then we must keep room for about the same number to prove our thesis.

Taking another example, let us suppose that we are going to use a noxa for which we know the $E.D._{84}$, and we think we have found a way of reducing its toxicity. As the preparation of the experiment is tedious and the substances deteriorate quickly, we wish to try and get a definite result in the first experiment, so that we aim at a result that is highly significant.

Under these circumstances we see where the 84 line in the vertical scale hits the quadrant and marks out the 84 per cent split. We then turn to Table 2 given by Mosteller and Tukey, and note that 10 mm. give a 95 per cent confidence level, 11·8 mm. a 2 per cent level, 13·1 a 1 per cent level, and 15·7 mm. a 0·2 per cent level for the usual double-sided significance. As we wish to have a highly significant result, we choose the 1 per cent level of 13·1 mm. and draw the line at this distance parallel and beneath the 84 per cent split.

We see from this line that 74 on the vertical scale just hits this line at the quadrant and therefore represents a percentage. From this we can say that if the treatment reduces the effect from 84 per cent to 74 per cent we would need 100 animals (74+26) to produce a highly significant result. The 99 per cent confidence line passes through the paired count of (50, 20) so that if the treatment reduced the effect to about 71 per cent. 70 animals would be sufficient. The line continues, to pass near the paired count of (15, 10), so that with an effectivity reduced to 60 per cent only 25 animals would be necessary.

Reading about binomial paper without using it, is a dull business, so that if you are interested in its possibilities you should buy a pad and play around with it. Its use in analysing data will be discussed later.

Numbers in mensuration experiments

Here again it is difficult to give specific advice that is realistic. Some help as to the number of animals that should be used can be had by looking at Fisher and Yates' tables for " t " (Table III). Looking under $p = 0{\cdot}05$, as the conventional limit for significance, we see that 3 animals that are necessary to produce 2 degrees of freedom, would need a distance of over 4 s.d.s before a fourth measurement could be classified as significant. This distance falls rapidly to about 2·6 with 5 degrees, and then more slowly from 2·5 at 6 degrees to 2·0 at 60. During this fall anything between 15 and 25 gives a limit of about 2·1.

When comparing two groups we shall be interested in the variability of the mean rather than that of the individual, and the standard error of the mean is calculated as $\text{S.E.} = \frac{\text{s.d.}}{\sqrt{n}}$ where n is the number of animals. Thus if you have $\frac{\text{s.d.}}{\sqrt{4}}$ and you wish to halve the standard error, you will need $\frac{\text{s.d.}}{\sqrt{16}}$, or 4 times the number of animals; to reduce the distance to half once again, you would require to have $\frac{\text{s.d.}}{\sqrt{64}}$ or 64 animals. Sixty-four animals where the " t " value is only 2, and where the s.d. will be divided by 8 to produce the standard error of the mean, is, therefore, a very good sample, but usually well beyond the financial limits of the experiment.

To generalise, even a small group should not have less than 5 or preferably 6 animals; a useful sample is one between 16 and 25, little is gained by having over 30 if the animals are fairly standardised; and a figure as high as 64 should not be necessary except when taking a random sample from a very variable population.

To calculate the number of animals required from the information that is available, is outside the scope of this book, but Snedecor (1940) gives a nomograph (Fig. 2.2) in which, by plotting the combined s.d. of the two groups and the difference between their means, one can read off the number of pairs necessary to obtain either a low or high level of significance. Alternatively Mosteller and Tukey (1949) demonstrate how, by using mean squares and a confidence zone, the degrees of freedom can be estimated graphically from binomial paper.

Surveys

The technique for carrying out a survey is outside the scope of this book, but an introduction to the subject may serve as a useful warning that it is a science as well as an art. A proper survey does not consist of a joy-ride round the pleasantest parts of the country, recorded as a collection of personal reminiscences, and concluded by a strong opinion based on the most vivid impressions. Like every other part of realistic research the survey is a job of hard work, carried out on well tried statistical principles.

The first action is to decide the purpose of the survey, for it is obvious that one survey cannot find out all about everything. Moderation must be used in the number of questions to be asked for once an attribute is selected for measurement it must be measured in every sample and there must be no empty gaps in the tables. Thus, if the survey concerns the nutrition of farm animals, for example, it is no good inspecting the pastures on 90 farms and neglecting the other 10 because on those occasions it was pouring with rain. Nor is it any good recording the weight of a diet and its analysis to two places of decimals at one farm and merely guessing at its contents at the next.

Questioning itself is an art, and one must try to avoid questions that can be answered by a simple " Yes " or " No." Thus the direct question " Are all your animals healthy? " demands the answer " Yes." whereas " When did you last have a sick sheep? " might open up the way to some information.

A great deal of care must be spent in deciding these questions for if an additional important question is only thought of halfway through the survey, then the whole of the first half must be re-sampled and the question asked, or the measurement made.

Number of animals to be sampled

Once again the size of the sample depends on whether we are dealing with enumeration or mensuration. Where measurements are being made any increase over a well-chosen sample of some 30 items merely varies the second decimal place of the 95 per cent confidence level, whether the population totals 100 or a million. What, therefore, we are probably looking for is the absence, or presence of unsuspected factors that will produce a bias in that particular environment, and thereby to split what appeared to be one population into different, smaller, populations. As these biases can be present, or absent, this would bring them under the heading of enumeration data.

The reliability of enumeration data would again depend on the size of the sample rather than the total size of the population. The number of samples to be taken can be judged to some extent by looking at the tables given by Mainland, Herrera and Sutcliffe. Here Table IX gives the number of A's in the sample as 0 and then records the percentage of A's that could actually be present in the population for different confidence levels.

Using the 99 per cent confidence level, and using the table the other way round, we can see that if we sampled 100 farms, or whatever the environment is, we would be unlikely to miss altogether a factor that was present in 4·7 per cent of the farms. Sampling 250 we would be unlikely to miss a factor present in only 2·09 per cent of farms, and sampling a well-chosen set of 500 farms we would have a 99 per cent confidence level that we would catch at least one farm showing a factor that occurred in only 1·05 per cent of them, or a 95 per cent confidence level that we would catch factors present in only 0·74 per cent of farms.

Choosing samples

In preceding paragraphs we used the term " a well-chosen sample" and this requires elaboration because it is the basis of a reliable survey. When carrying out a survey local workers may advocate certain places for personal reasons. A visit to one locality may mean an invitation to lunch; a locality far away may bring extra travelling expenses; and in a third locality obtaining the required samples may be easy and convenient. All these temptations must be put aside for the golden rule of sampling is that every animal in the population has the same chance of being sampled.

Where sampling is being done on an area basis then a large area can be broken up into squares, and certain of these squares can be selected by random numbers. If necessary these squares can again be sub-divided. As in experimental work, to ensure that you get probable sources of bias accurately represented areas can be stratified into, say, high and low districts. If desired these areas can be sub-stratified into those inland and those near the coast. This sub-stratification can be continued, as long as it appears practical, until you feel that the important sources of bias have been included.

Once the areas have been allocated, the size of the sample to be taken must be decided. If wild animals are being dealt with this could be a fixed number irrespective of age or sex, but with domestic animals that are readily available stratification will usually be employed. For example, with cows, the various stratifications might be young or old, pregnant or non-pregnant, pastured or housed, and one breed or another; animals being drawn for each class by using random numbers.

Where a survey is likely to be analysed on the basis of a factorial experiment one should not place the number to be sampled in each area too high, for if in some areas they are not available it will spoil the structure. On the other hand where local conditions are of special interest there is no reason why a large number of animals should not be sampled even if some must be omitted for part of the calculations.

Drawing conclusions from a survey

A survey is employed to collect information. It shows the range of measurements and their variation under different sets of circumstances. In general, however, it is not a reliable method of proving cause and effect. A different environment often covers several variables so that, to repeat a previous warning, you cannot compare a flock of sheep of one genetical line, at one altitude, on one type of soil growing one type of herbage, with another flock in a different environment and say that any physiological difference in the blood is due to a difference in breeds. Nor can you, even when a χ^2 test has shown a significant association between a disease in animals and their infestation with ticks, say that the ticks have caused the disease, or even carried the disease, for the environment in which the disease can spread may also be the one that is suitable for the existence of ticks; further, if little is known about the ecology, it might merely indicate that diseased animals were likely to become infested.

Further reading

These introductory remarks on carrying out a survey are a warning that there is a correct method and that this should be followed. Before attempting a survey one should consult *Sampling Methods for Census and Surveys* by Yates. This eminent statistician has kept the first half of the book free from complex mathematics so that it will serve as a guide to non-specialists. For the non-mathematician there is also a lot of information in the book by Wallis and Roberts, or if you wish for a medical bias there is much information in Bradford Hill's little book.

Summary

In allocating animals to experiments, no animal should be included unless you are prepared to accept the information it will provide. Including abnormal animals, just to make up the numbers, may produce queer results. Where biases are associated with detectable qualities, animals having these different qualities can be separated into sub-treatments of proportional sizes providing their own sum of squares, they can be distributed as equally as possible among the treatments by using restraints, or they can be randomised among the treatments without any special handling. Manipulations after allotment must be avoided, or be very few in number and even then carried out with an uneasy conscience. The final act must always be the random selection of which group receives which treatment.

Surveys are not joy-rides round the country in which a few samples are gathered and people are asked any question that comes to mind. They are a piece of hard work based on accepted rules, and even then the results must be interpreted with care as to their logic.

References

MAINLAND, D. (1938). "The Treatment of Clinical and Laboratory Data." Oliver and Boyd (London), Section 12, page 305.

—— (1952). "Elementary Medical Statistics." W. B. Saunders Co. (London), page 101 and page 268.

MOSTELLER, F., and TUKEY, J. W. (1949). "The Uses and Usefulness of Binomial Probability Paper." *J. Amer. Statist. Ass.*, **44**, 174.

SNEDECOR, G. W. (1940). "Statistical Methods." Iowa State Press (Iowa) 3rd. Edn., Section 4.3 page 70.

CHAPTER 12

COLLECTING AND RECORDING DATA

> Gentlemen, the most important part of living is not living but pondering upon it. And the most important part of experimentation is not doing the experiment but making notes, very accurate quantitative notes—in ink. I am told that a great many clever people feel they can keep notes in their heads. I have often observed with pleasure that such persons do not have heads in which to keep their notes.
>
> SINCLAIR LEWIS (" Martin Arrowsmith ")

Having designed an experiment that will give a clear " Yes " or " No " to the hypothesis, and having dug the foundations so that it does not topple over, the next thing to do is to get on with the building. To do this we need bricks that consist of measurements, and the skill to place them neatly in the correct place. No worthwhile structure can be built without these well-paced bricks that form the data. The neat and accurate collection of data is an integral part of a reliable experiment, and results are untrustworthy if they are based on the reading of half obliterated numbers on the sample tubes, or are recorded with a blunt pencil, or a ball-point pen, on the backs of envelopes or on empty cigarette packets.

COLLECTION OF DATA AS A CHECK ON THE EXPERIMENT

When we split an investigation into its parts, the collection of data was the third action of the series, but the collection of data is not restricted to this position and fresh opportunities for collection occur with every experiment, providing clues to other hypotheses and helping the worker to check his experiment. When we speak of a " well-controlled " experiment we do not refer to the presence of control groups alone, but include this check given by the data.

These selected serial measurements, that constitute the collection of data, may uncover the unexpected abnormal value, and this may be the most important observation in the whole experiment.

Sometimes these aberrations point the way to another theory but at other times they reveal some unsuspected error. To stress the importance of this aspect I would like to give a few examples based on field experiments, for these are easier to understand, but a little reflection will show that similar errors can creep into laboratory work.

For the first example let us suppose that we are immunising one group of animals by giving a course of vaccination, and leaving a similar group untouched until finally we give both groups a test dose of the organism that is being investigated. In most animals coming in contact with these organisms antibodies are produced and their presence can be shown by a blood agglutination test, but as it is known that (i) the titre of the agglutination bears no relationship to the immunity of the animal, and (ii) that the control group will remain negative throughout. Why bother to do the test?

The answer is that if nothing goes wrong we will have gained nothing but extra confidence in doing this test, but, something might go wrong. If there is an absence of agglutinins in one of the vaccinated animals there is the possibility that it has been overlooked when vaccinating. If, in addition, agglutinins appear in an animal in the control group, then there is a chance that the numbers have been mixed or that one cow is in the wrong position. Finally, if agglutinins appear in the control group and investigation shows that none of the group has received even a single injection of vaccine, then there is the possibility that they have been infected from some natural reservoir of the disease. You may protest that it is an insult to suggest that you might vaccinate the wrong animals and that it is ludicrous to suggest that the controls could become naturally infected and these protestations might be quite true, but you will find it very reassuring to have completely negative figures for the agglutinins in the control animals when chance does decide to give you a very peculiar answer. By keeping these measurements you are pointing the gun at your own head in trying to prove yourself wrong, and it will be these measurements, and these alone, that will give other people enough faith to accept the peculiar results.

For another example take a field experiment in which you are supplying a growth-promoting factor and in view of the warnings given in Chapter 2 you are administering it yourself. This factor is detectable in the blood and faeces. The experiment is to feed this factor to one or two comparable groups of growing animals and then compare the growth rates. If you are always going to feed the animals yourself why bother to take blood or faeces samples, particularly in the control group? You will be there yourself to see if anything goes wrong, and you cannot be expected to make a really silly mistake. The answer is that if you are going to make a claim and expect other people to believe it, then you must show that you have taken precautions against mistakes. Imagine that it was an iodine deficient area and the active principle of your compound was iodine. If you were doing the experiment correctly and were testing the levels in weekly samples you might find that one week some of the control animals showed higher levels of iodine than those receiving it. Question, and perhaps accusations, might follow, but in the end it might turn out that the attendant had quite legitimately dosed the animal on the advice of his veterinary surgeon, or that he had been treating some of the control group for ringworm by painting the lesions with a preparation containing iodine.

We have supplied two examples without questioning anybody's ethics but one must also be prepared for errors occurring through self-interest on someone's part. Hence, as a final example let us take an experiment in which certain herds, infected with a form of mastitis, are treated with an antibiotic which can only be obtained by prescription. The plan of this experiment is that milk samples are sent in each week to a laboratory. Where the infecting organism is found in the sample, a full treatment of antibiotic is sent back to the owner for use on that animal, and he is asked to send a series of daily samples. If we wished we could examine all samples, not only for the presence of the infecting organism, but for the number of leucocytes, for clots, and for the presence of the antibiotic. But why waste time on this? All we are interested in is to see how quickly and constantly the organism disappears once the antibiotic treatment has begun, and all the owner is interested in is getting his cows free from mastitis.

Let us answer this question by looking at things from the owner's point of view. He finds that the antibiotic is effective, in fact, his belief is so great that he would like to possess a little for his own use, either in the future or for cows in another herd, and he knows that the laboratory has got plenty of the stuff or they wouldn't give it away in experiments. At the next test, therefore, the owner finds clots in the milk of cow number 49, so that, as well as putting a sample from this cow into bottle 49, he also puts it into bottles 7, 12 and 63. The report comes back: cows 7, 12, 49 and 63 are infected, and this is accompanied by the appropriate amount of antibiotic and the request for a series of samples. The owner has now obtained his three spare treatments and for the time being is satisfied with this, so he now puts the correct samples into the correct bottles. The laboratory may be delighted, for, by measuring infection alone, the serial samples will show 3 cows in which there is a total and permanent absence of the organism after only one dose of antibiotic. A more careful laboratory that recorded not only the presence of infection but the strength of the antibiotic in the milk, the presence of clots and the numbers of leucocytes would be in a different position. They would see the absence of the organism without any antibiotic being present and the sudden change in clinical tests from acute mastitis to perfect normality; discrepancies that would demand investigation.

Three examples have been given to show the value of collecting data as a control on the experiment, we have seen how in addition to acting as a guard against gross errors on your own part, they help to protect against external interference, in one instance interference from a natural source, in the second from an innocent attendant and in the third a less innocent one. Apart from these differences the examples are alike in the facts that: (i) they are all based on fact; (ii) the interference could cut right across the logic of the experiment and lead the observer in the wrong direction; (iii) they were all discovered by carrying out measurements that were quite unnecessary for the primary object of the experiment; and (iv) the errors could not have been demonstrated in any other way. These examples show what an advantage it can be to point the gun at your own head by the collection of data and underline a fundamental of experimental work, that it is far better to find out your own mistakes than to allow other people to discover them.

Collection of data as part of the experiment

In planning an experiment we plan to get a conclusive answer to the primary question, we introduce any secondary question, such as, does the interference produce the same result in young as in old animals? And we arrange the recording of the data necessary for answering these questions. This gives us a nice little experiment that could be done as an enjoyable hobby, leaving plenty of time for reading, lecturing, discussion and more enjoyable occupations. If, however, we are full-time research workers we should not be content with this easy approach and in view of what has been said we should turn it into a professional job by an additional collection of data. From the ethical point of view this collection of data is even more important when the experiment involves pain or discomfort to the animals involved, for, however ardent your belief that experiments on animals are essential to man's survival, no one can defend the position where, in one room an animal is subjected to an interference in order to study the effect on the urine and kidneys, while in another room another animal is subjected to the same experiment so that someone can examine the blood changes. Experiments of this sort demand team work from the ethical and financial standpoints, and in addition they demand it from the standpoint of efficiency, for it is only when several measurements are made together that the unexpected correlations between one measurement and another, or the dramatic discrepancy between measurements, can point out the essential dynamics of the physiological process within the animal and raise the hem of the curtain to give some inkling of the interaction of events.

For the collection of data one can choose appropriate measurements from the whole repertoire of clinical pathological or physiological tests, beginning with the pulse rate, respiration rate and body temperature; measurements of gains in weight and food consumption; analysis, cytology and bacteriology of secretions and excretions; biopsy examinations of blood, liver and marrow: and finally post-mortem examination with its appropriate histopathology. Such an investigation may require team work with a pathologist, a bacteriologist and a biochemist, but at the end of the experiment instead of having merely answered the question

"Does A produce B?" you will know a good deal about the general reaction of the body to interference.

To recapitulate; If you take a nice little experiment with its primary and secondary questions, and then load it with the collection of appropriate data until, instead of being fun, it becomes hard work, then you will have changed from a dilettante to a realistic research worker.

Pre-experimental samples

When it is possible, it is usual to take some samples before introducing the experimental interference, either to establish a level for the starting point of a graph, or for later use in an analysis of co-variance.

Greig and Boyne (1956) working on the haematology of calves state that if the variation in the daily samples of an individual animal are small as compared to the difference between one animal and another, then efficiency will not be increased to any great extent by having numerous pre-experimental measurements on each animal. Under these circumstances three preliminary estimations are suggested as enough. This routine of three measurements is a practical one and would probably apply to most physiological experiments, but it is worth noting that the chance of three measurements falling in any specified order is $\frac{1}{3 \times 2 \times 1}$, so that wherever measurements vary in a chance manner then once in every 6 times you will have the three measurements in an ascending order (e.g. 5·3, 5·7 and 5·9) and in a descending order (5·9, 5·7 and 5·3). Thus 1 animal in 3 may show an upward or downward trend and if this trend lies in the same direction as the one you expect from your experimental interference you will feel less confidence in the result. For this reason if the animals are few, five preliminary tests may give a more convincing preliminary.

These authors go on to say that where you are dealing with such animals as monozygous twins, using one member as the experimental animal and the other as the control, then here the error in technique, plus the physiological variation in the animal, is large when compared to the difference between one twin and the other, and here a much larger number of preliminary samples will be practical as this will permit the averages of the pair to be fixed within a close range. As monozygous twins are only of value in detecting small changes requiring accurate measurements this accuracy is important. As an indication of the increased efficiency the authors give the following figures. One sample gave 31 per cent efficiency, 3 gave 58 per cent, 6 gave 73 per cent, 10 gave 82 per cent and 20 gave 90 per cent.

With regard to routine experimental samples during an experiment, the rule here is that the more inaccurate the method of measurement, the larger the number of samples to be taken. With inaccurate measurements 12 samples, or multiples of 12, are elastic, for they permit 6 averages of 2 measurements, or if these results are still too erratic to show a trend, then averages of 3, 4, or even 6, can be tried. Thus when planning an experiment it is as well to give a thought to the possible averages that can be made from a series and to avoid a figure like 7 for the total number of tests.

When dealing with a regression it may be remembered that with the method of least squares it is the items at the extremities that add the greatest quantity to the sum of products, this means that it is these extreme values that set the slope of the regression line. Thus to get the correct slope several samples are required at each extremity rather than an even distribution over the whole line. Where samples are taken over a period in which the interval between samples increases as a geometrical progression, it may not be illogical, therefore, to end the series with 3 or 4 samples bunched together. Where conditions involve the use of probits the samples at each extremity should be increased even further owing to the unreliability of the extreme values in these conditions.

Duration of experiment and frequency of sampling

It is difficult to give more than general advice on the length of an experiment for this depends so much upon the condition being investigated, thus anaphylactic conditions begin in a few seconds, may be over in a minute or two and the body back to normal in about two days, in contrast carcinogenic substances may produce no visible effect for several months followed by the sudden appearance of a malignant growth, while with the experimental transmission of scrapie, symptoms may not be visible until the end of two years. With milk yields in cows the possibility of the udder compensating for a temporary loss by a subsequent increased production must be considered, so that sampling may be continued after the experimental interference is ended.

The only help that can be given is to suggest that where you are dealing with the unknown, you work on the hypothesis that acute things happen quickly and changes that come late usually come slowly. On this hypothesis the times of sampling in unknown conditions can be based on some general geometric (logarithmic) series. Such a series may often be a loose mixture of arithmetic and geometric sampling dependent on the previous experience of the worker.

An example of a rough geometric series is as follows: Monday spent in preparing the materials and in taking a preliminary sample. Tuesday, taking the final preliminary sample, carrying out the experimental interference, and sampling at 5, 10, 15, 25 and 45 minutes, followed by samples at 1, 1·5, 3, 6 and possibly 9 and 18 hours. This is followed by morning and evening samples on Wednesday, and morning samples on Thursday, Friday and Saturday. In the second week sampling on Monday and Tuesday followed by Thursday and Friday; a poor series but capable of providing 2 averages of 2 measurements for that week. A sample taken on Monday and Friday in the third and fourth weeks and then falling to once a week as long as necessary.

This series is not offered as a practical one, for usually enough is known of the alterations likely to occur to permit the series to be cut at one end or the other, but parts of it can be used as a rough guide, and extra samples can be interpolated or alternate samples omitted as appears practical.

Anaphylactic shock will need samples taken at seconds as well as minutes, substances causing neutrophilia and inflammatory changes will need the hour series as well as the daily sampling, but where infections cause a transient rise in temperature, no geometrical series is possible and temperature must be taken at least each 6 hours, day and night, Saturday and Sunday included. Note that in the rough geometric series Sunday sampling was avoided. This is not because the comfort of the investigator has been studied, but because Sunday, with the laboratory stores closed, transport difficult, and animal attendants eager to be off, is a day of low efficiency.

Two other factors concerning a time series are as follows: Firstly, if in the unemotional atmosphere of planning an experiment you set out a time series, then keep to it; do not in the emotional upset of an unexpected result start cutting the series unless you are absolutely certain that the full sampling would be a waste of time. Secondly, when you plan the series in the first place, remember that the longer it runs the more unknown factors will play a part, thus the conditions of the experiment may lead to a weakened resistance to disease in both groups, and which group suffers most from these intercurrent infections may depend on chance. Thirdly, if you get an unusual measurement, check it by taking another sample as soon as you can, even if the plan says that the next sample is in the next week.

Sampling

Each animal to be sampled must be easily identified by such methods as: tattooed numbers; brands on the hide, hoof, or horn; ear tags, punched or clipped ears; or a collar with a numbered disc. Identifying by fixing a card to a cage, or pen, is not fool-proof, for when cages are being cleaned the cards may be knocked off and returned to the wrong cage. If in one cage in the experimental group the reaction is similar to the controls, and one cage in the controls is similar to an experimental group, then in some circumstances this could lead to an exciting new hypothesis, but here one must remember Occam's razor. Occam said choose the simplest hypothesis that will fit the facts, and in these conditions the simplest hypothesis is that the cards have been interchanged and, unless you are absolutely certain that this is not the case, you can go no farther.

Not only the animal but the sample taken from it must be clearly numbered and dated. In rough field conditions the tubes can be numbered in the cleanliness of the laboratory and the appropriate tube chosen when the animal is to be sampled. Here again, it is better to number the tube itself rather than the stopper, or the position in a rack, and so prevent interchange. In wet conditions it is preferable to label tubes with surgical plaster, or to cover gummed labels with transparent tape.

Need for absolute units

Sometimes in addition to taking samples, a diary of clinical observations is kept. Every effort must be made to keep this absolute rather than relative, and whenever possible the differences should be graded into numerical units. It is very easy in recording clinical appearance to use terms such as " Off colour " one day, " Listless " another day, and " Not well " and " Dejected " on other days without any clear idea of why a different phrase was used. In examining records later it is difficult to assess how many daily reports of " improved " or " slightly better " are required to turn a " Very dejected " animal into a " Recovered " one. If possible, therefore, record as much as possible in a factual manner. Does the movement of the ears indicate an interest in what is happening? Do the eyes follow your movements? Is the head held normally or allowed to hang? Does the animal try to avoid capture, or resent manipulation? As these signs occur record them as present, and thereafter record their presence or absence, preferably by turning the observations into the columns of a table.

Similarly with measurements, if size is important, do not record an egg sized swelling but measure it. As an inset to volume 44 of *The Journal of Pathology and Bacteriology*, there is a delightful essay on the " Quantitative Study of Tumours." This essay suggests that if eggs are really going to be used to measure the size of tumours then standard eggs should be deposited with the Royal Mint or with the National Physical Laboratory. Nevertheless, if it is impractical to measure an object accurately, then even the comparison to the size of a hen's egg, or to a coconut, is much more definite than to talk of a " small " or " fair-sized " object, and with a little ingenuity a crude scale can be set up for most qualities. Such a scale can be accurate enough to permit the results to be put into an array so that a non-parametric significance test can be carried out later to give an unprejudiced conclusion.

Do not despise these crude measurements because they may be simple. The virtue of simplicity was illustrated when one of our tractor drivers was told that a surveyor had measured the area of a field by means of a new instrument, and the area was

35.63 acres. " That's not right," he said, " It takes me just three days to plough that field, it will be just about 22 acres." And of course it was, for the simpler the method the more difficult it is to make a really big mistake.

Rough recording

When making measurements always try to cut down the risk of errors by following a set routine, arranging the sample tubes in the same order, and beginning at the same place. Record the actual measurements in a reporter's notebook, or an exercise book, with the pages ruled in columns and rows so that any omission is immediately obvious. Record the actual reading; turning it into a percentage, subtracting the weight of the bucket, multiplying by a correction factor, and other adjustments can be left until the results are entered into the permanent records. If possible keep the actual samples until the figures have been entered in these permanent records and are seen to be congruous; this allows the chance of re-estimating an unexpected figure. The whole structure of the experiment depends on these measurements, or observations, so write them down clearly so that their correct interpretation is not a matter for debate.

Permanent records

After even ten years of research work you will have built up a lot of records, and it will depend entirely on you whether this data will still have value, or whether it will represent a collection of various sized notebooks filled with meaningless figures. In case you think that I am over-emphasising this danger I will quote from Cochran and Cox (1953). They say, " Much data, expensive to collect and potentially valuable for some research, have had to be discarded or destroyed because the original collector cannot be reached, and new investigators are unable to discover what measurements were taken and under what circumstances." I would go further and say that, as the original collector did not make a note at the time, even if they did find him, he would not have the slightest idea what it was all about.

Permanent records should be so complete that they will permit you, or even other people, to understand what the experiment was about, even if consulted after a lapse of years. Records should therefore include a plan of the experiment, the materials and methods, and the basic tables recording the experimental results.

If one records in exercise books then one experiment may fill but part of a book whereas the next may require a few pages more than the book contains, so that for permanent records loose leaf books are to be preferred. One good method is to use strong, good quality loose leaf binders that hold sheets punched with 4 holes and measuring $12\frac{3}{4}'' \times 9''$. These binders hold about 150 sheets and when the experiment is completed the records can be removed and placed in a cheaper, larger, transfer binder that holds about three times as much. The actual sheets can include plain sheets for drawings or typewritten work, sheets ruled with lines, squared paper with rulings of 4, 5, or 10 to the inch, sheets with columns, and those with special use. For example, there may be forms giving a temperature chart for a period of a month, post-mortem sheets detailing the organs, and other special subjects such as the results for blood examinations over a series of days.

A punch should be available so that correspondence, and other types of paper, such as semi-logarithmic paper, can be inserted, or a punched polythene bag holding photographs, or possibly card-index cards, can be introduced. A series of these binders can be held upright by fitting racks to a laboratory cupboard.

General plan of the experiment

The general plan of the experiment is usually well discussed with one's co-workers as part of the preliminary work, but it is seldom committed to paper. In a short time one can easily forget which worker was responsible for which piece of work, or which of three methods was used, so there is ample reason for recording such details and one method is to record them along the lines of army operation order, as follows:

Information. A brief note of the observations that gave rise to the hypothesis.

Intention. The primary object of the experiment.

Method. The experimental interference that is to be carried out in the different groups.

Administration. Recording which worker is carrying out which part of the experiment and, if necessary, the rota for week-end duties.

Intercommunication. The amounts of materials and reagents that will be required from other departments, such as the animal house, the media kitchen, or the storemen.

Recording this information requires self-discipline which is unpleasant, particularly to those who feel that their imagination borders on genius, but it is a necessary part of efficient research work. Although it may seem over-conscientious to record even the reason for the experiment it is surprising how many senior research workers, presented with old records, ask " Whatever did I do that for?" often in less polite language.

Recording data

The experimental records can begin with a sheet or two for use as a diary of events. This can record the times of treatment in the different groups and any event that might affect the experiment, such as a change of diet or a change in animal attendant.

Temperature charts form useful record sheets for animals where clinical temperatures are being recorded, or where the sheets can be modified to record some other measurement such as milk yield. Each sheet is clearly marked with the number of the animal and arrows can be marked in to indicate the dates of treatment, the actual treatment being written out in detail over the graph if it varies from animal to animal. Clinical symptoms can also be recorded on this sheet, and if the animal dies, and a special form is not being used, the post-mortem findings can be recorded on the back of the sheet. By this means the complete history of each animal is kept on the same sheet.

Sheets for serial samples are headed with the name of the attribute to be measured, followed by a note as to what the figures mean, even if at the time this seems glaringly obvious. For example, if a sheet headed " Milk Yield " bears figures such as 7·3, does this represent litres or pounds?

The identifying numbers are usually in the left hand column and additional columns record the dates at which samples were measured. Actual readings are recorded for these will be used in statistical tests, and the conventional values calculated from these figures can be shown in an adjacent column or even on another page. The method of calculating the transformation must be shown on the first sheet. Thus if one were calculating the number of organisms, or leucocytes in milk, the heading might show the amount of milk in a film and that the figures represented the number of organisms counted in 12 fields of microscope No. 172 using No. 628 2mm. lens and a times 8 eyepiece. If for convenience a better microscope with different lenses is used later, this detail must be recorded for it will lead to a different calculation in obtaining the conventional value. The adjacent column might be used for the number of organisms per millilitre, and the calculation necessary to achieve this must be recorded. It is surprisingly easy to forget to record this kind of data with the possible result that if you suspect a mistake in the calculation you cannot correct it, for a year later you have forgotten the formula you used, or even how you came by it.

In recording, avoid what Moroney calls the "delusion of accuracy." If, to get a conventional standard, you add 00 and divide by 6, then record 34 as 566 and not 566·66. The number of significant figures you can usefully record will be obvious from the results of the duplicate tests you made when testing the technique.

Again let me emphasise that for ease in an analysis, ease in writing the scientific paper, and for certainty in drawing the conclusion, it is most important that each space in the table is filled. Any empty space will be an embarrassment, for these tables will go to form the basic table in the resultant scientific paper that records the experiment, and any blank space filled with a contrived figure is merely the expression of your opinion.

Summary

The various measurements that are made during the experiment represent the whole brickwork of the structure. These measurements may act as a further control on the actuality of experimental interference and form the source of fresh hypotheses; they also represent the result of the experiment. The importance of these measurements demands that they should be free from errors as to which animals they come from, and that they should be recorded in a way that is easily decipherable. Because accurate data always retains a potential interest records may be consulted long after the interest in the actual experimental interference has been lost. For this reason an explanation of the techniques used, and the units in which the measurements were recorded, should be included in the records.

References

ANON (1937). "On the Quantitative Study of Tumours." *Jour. Path. Bact.* **44**, Insert.

COCHRAN, W. G., and COX, G. M. (1950) "Experimental Designs." John Wiley and Sons, (Chapman and Hall) New York. Section 3.34, page 55.

GREIG, W. A., and BOYNE, A. W. (1956). "The Effect of High and Low Planes of Nutrition on the Haematology of Monozygous Twin Calves." *Jour. Agric. Sci.*, **47**, 150.

CHAPTER 13

INSPECTING AND ANALYSING THE RESULTS

It has been alleged that certain people use statistics as a drunk does a lamp-post—more for support than illumination.

CHURCHILL EISENHART and PERRY W. WILSON

When the basic tables have been completed, and the experiment is at an end, the time has come to examine the figures to see what they mean. Here we must remember the warning about " not seeing the wood for the trees." Some people have the ability to look at the figures and realise the pattern they make, but most of us need to gain a better perspective by using statistical methods. To do this, instead of sitting down and staring hard at the figures, hoping for sudden inspiration, we start work with a pencil and paper.

If the data can be classified into different categories or ranks we can start working out frequency tables for each of the groups. If the data consists of measurements then we can make dot diagrams. To do this we look for the minimum and maximum figures in the measurements we are dealing with, and then roughly scale a row of squares to cover this range. We then put a dot in the middle of a square at the appropriate measurement for each item and thus build a rough histogram that will indicate the type of distribution that we are dealing with. Beneath one dot diagram we can build another for one of the other groups, and this may well show that the two samples come from different populations, alternatively, it may show either that they are both samples from the same population or that you need the help of a significance test before you pass judgement.

Where there is a possibility that two things (two variables) such as height and weight, are associated, then a rough scatter diagram can be made in which the height might be shown by the horizontal scale and the weight by the vertical scale, the approximate measurement being dotted into the appropriate position. If you have three variables you could use something similar to a draughts-board (or chequer board) to give you horizontal length and breadth scales and introduce the third scale by the difference in height of piles of coins placed in the appropriate positions.

These rough manipulations are an important part of the analysis, the measurements being arranged, and re-arranged, to answer different questions. The common methods given here should cover most results but if they do not it is well worth while spending time working out some other simple method of depicting the pattern of the measurements at the expense of the individual accuracy. This first inspection is followed, where necessary, by tests for significance which are dealt with below.

Analysing contingency data

Contingency data is another term for enumeration data, and refers to things happening or not happening. The data to be used in demonstrating this type of analysis may be the proportions in which experimental animals of different groups lived or died. Contingency tests, however, are not necessarily restricted to the presence or absence of different qualities but are often used to obtain a quick, objective judgement on groups of measurements. This is done by making one particular measurement the criterion as one does in grading hens' eggs as above, or below, 2 oz. In the same way the measurement in a benign disease might be based on the weight of the spleen—the heavier the infection the greater the weight of the spleen—and the frequencies could refer to weights above the average spleen weight and those below it, alternatively they might refer to plate counts graded as above, or below, 10,000 bacterial colonies per gramme of spleen, if that was the organ selected. In this way we are using contingency tests rather in the same way as one might tighten a nut with a pair of pliers. If one could tighten it sufficiently by this method there would be no need to fetch the correct size in spanners, while in those cases in which one failed to get the nut tight enough it is still possible to try again, using a spanner, which in these circumstances would be analogous to a ranking test or to the analysis of variance.

Binomial paper is also of use as a quick survey method, but when used for a four-cell test where the split is an empirical one based on the average for the two groups, we must shorten the perpendicular distances by multiplying them by 0·7. As a hint to those who may wonder where this figure comes from, 0·7 is the reciprocal of $\sqrt{2}$ which enters the formula for the standard error of the difference.

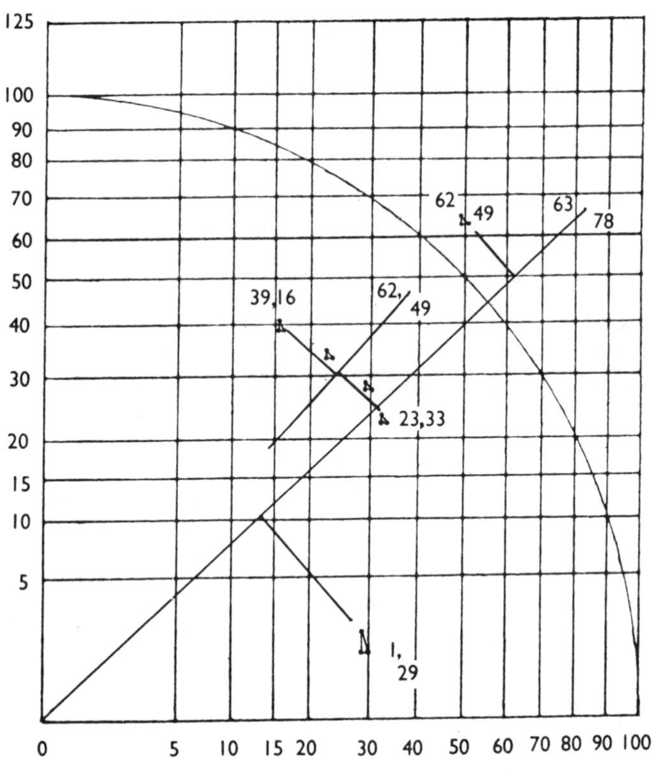

FIG. 240. Use of binomial probability paper to test the significance of 2×2 tables. Note that, as the scale for standard deviations is omitted, the distance between 90 and 100 can be used to measure 1 s.d.

Taking as a simple example the following 4 cell table:

	Pass	Fail
Group A	62	49
Group B	1	29
	63	78

we can see from Fig. 240 that the empirical split has been drawn to the point 63:78, and one paired count gives the triangle (62, 49) (63, 49) (62, 50) that falls near the top of the split, and the other (1, 29) (2, 29) (1, 30) falls near the bottom of the split. As there is no scale of s.d.s. in the part of the chart shown, we can use the distance between 90 and 100 as equal to 1 s.d., this gives the perpendicular distances to the split as $4{\cdot}2+2$ s.d. and converting by multiplying by 0·7 gives 4·34 s.d. This is very highly significant, being off the table for the normal curve and well outside the limit given by the last row in the *t* Table in the appendix.

With, for instance, a 2×2 factorial design, binomial paper offers a very quick test because the Total split remains the same for all comparisons. As an example take the following Table that shows the results of comparing a live and dead vaccine at both larger and small doses, in an experiment that began with 28 guinea pigs in each cell.

	Live Vaccine	Dead Vaccine	
Large Dose	12:16	22:06	34:22
Small Dose	11:17	17:10	28:27
	23:33	39:16	62:49

From the Table we see that the split is based on 62 healthy animals and 49 sick ones. Entering this on Fig. 240 we can then plot a comparison between the results for a live vaccine in comparison with a dead one, that is, 23:33 against 39:16. The result amounts to $(2+1{\cdot}6)\times 0{\cdot}7$ giving 2·52 s.d. a value that the *t* table in the appendix gives as $p=0{\cdot}02/0{\cdot}01$; a significant result.

Comparing a large dose with a small one we have 34:22 against 28:27, these are shown in the figure as triangles but the perpendiculars to the split have been omitted. The total distance amounts to about 1 s.d. and is not significant.

Finally, comparing the interaction, we have 12:16+17:10 against 22:6+11:17 giving 29:26 against 33:23. These paired points fall even closer together and obviously do not portray significance.

Poisson distribution

This distribution was mentioned in Chapter 7. Its characteristics are that there is a large bulk of material in which the chance of the event occurring is small. As an example we might take the number of punctures a motorist has in a year, or each 10,000 miles; the event of a puncture occupying, perhaps half a second or, perhaps a few inches of road, it is difficult to say. If you averaged 3 or 4 punctures a year then a histogram for 30 years would give a skewed binomial-like distribution, and the χ^2 test would be applicable, but where you averaged less than one puncture a year the distribution would give a *J* curve, with the mode at nought.

Where the data suggests a Poisson distribution then most biological data can be dealt with sufficiently accurately by using Poisson paper. With this paper the average frequency of the event is plotted as " a " on the base line. The frequency of the event occurring in different samples is shown as " c " in a series of curved lines, and where these frequencies intersect the average frequency of the event we can read the likelihood of the event occurring " at least c times " from the scale of probability on the left hand side. If you wish to know more about this paper you will find simple examples of its use given by Moroney.

Measurements

Measurements and frequencies are to a certain extent interchangeable. It has been shown that sets of measurements may be coded by taking the frequencies of, for example, measurements between 0 and 100, between 100 and 200, and so on, and then comparing the different sets by the χ^2 test. The converse of this is to take frequencies and then turn them into measurements by calling them percentages. By recording these percentages as the sine of the square root, as seen on the binomial paper, it is often possible to get a normal distribution that will permit an analysis of variance to be made. This arcsin transformation, and others, are described by Snedecor.

Where measurements are only about 12 in number in each group it is worth while trying Wilcoxon's or White's test (Chapter 7 and Snedecor) as a test for significance, for if one of these shows obvious significance, or obvious non-significance, it will not be worth while going any farther. If, however, the result verges on significance it might be worth while to try an analysis of variance. In circumstances where there are more than, say, 12 in each group the tedium of ranking the data may in any case appear greater than that of an analysis of variance, particularly if a calculating machine is available. If data is extensive then time is saved by grouping the measurements into different classes. In making the calculations, as with histograms, the measurements used in the calculations are the centre values of each class, and where a measurement falls exactly between two classes each class can be credited with half a unit.

Correlations

To get the run of a line it was suggested earlier that you should make a scatter diagram of the approximate measurements, that is, measurements in which the last one or two significant figures have been dropped off. With a large number of items even this can be tedious; under these conditions one can pick out some of the lowest values, some of the highest values, and some average values of one measurement, and see how the average values of the associated measurements of these 3 groups compare. If they give a horizontal line it is obviously not worth going farther. Remember the warning by Wallis and Roberts given in Chapter 7, and also the statement there that if your figures average to 5 ascending steps you have demonstrated a trend. If the ascent is not regular then you may still be able to prove a trend quickly by using a ranking test, or even ranking your data and carrying out a routine correlation test.

Note that if you are comparing actual measurements with a theoretical line then by chance alone half the items should fall above the line and half below, so that you can test the goodness of fit with a χ^2 test or binomial paper. Remember also, that if you put a horizontal line through the average of one measurement and a vertical line through the average of the other, you can test your dot diagram with a four-cell χ^2 test or, easier still, carry out the quadrant test of Olmstead and Tukey (see Chap. 7).

Sometimes the different items jump about so much that it is difficult to appreciate the existence of a general trend. In these circumstances one may use a three- or five-point moving average. To illustrate a three-point moving average let us work one out for the first 5 natural numbers. The first average would be from $1+2+3$, giving 2, the next $2+3+4$, giving 3, and the next $3+4+5$, giving 4. Note that moving averages do not affect a straight line but they accentuate concave curves and diminish convex ones.

Adding probabilities

Sometimes one does an experiment and fails to get a decisive answer. Then, according to the strength of one's faith, one either gives up the idea or tries again, in the hope that the replicate together with the original answer will reach the level of significance. One's first thought is that if the first answer showed a level of $p = 0{\cdot}2$, or $\frac{1}{5}$, and the second one the same, then the probability would be $\frac{1}{5} \times \frac{1}{5}$, or $p = 0{\cdot}04$ and that this would be significant. Unfortunately, the correct method of combining probabilities is neither so easy nor so optimistic.

One way of combining probabilities is easy; that is when they are recorded as χ^2 values. Let us assume that we have a hormone that we hope will encourage the production of twins. Our first experiment on the lamb crop of ewes has given a χ^2 value of 3 when comparing treated and control groups in a four-cell test. $\chi^2 = 3$, with 1 degree of freedom has a probability of $0{\cdot}08$, and is not significant. Our second experiment is on goats and the resultant kid crop. This also gives $\chi^2 = 3$. We can now ask the question: On this combined evidence can it be said that we have produced a bias? To find this out we add the 2 χ^2 values and look up the result for 2 degrees of freedom. Under this heading we find $p = 0{\cdot}05$ gives $\chi^2 = 5{\cdot}99$, hence our result of 6 is significant. Note that, of course, the bias must be in the same direction in all the experiments you add together in this way, and that the level of average homogeneity is $p = 0{\cdot}5$ (similar to a coin tossed once) so that if you add χ^2 values that are near $p = 0{\cdot}5$ the total becomes even less significant as the degrees of freedom increase.

The routine method of adding probabilities can be applied to the results from any significance test, it is described by Mainland (1948), by Snedecor (1957) and, originally, by Fisher (1938).

PRECISE PROBABILITY OF INDIVIDUAL RESULTS

Many tables of probabilities are limited to selected levels, such as 0·05, 0·02 and 0·01 or 0·05, 0·01 and 0·001. For combining probabilities, as mentioned above, it is necessary to know the probability corresponding to the value obtained from the test, and not just that it lies somewhere between 0·1 and 0·05. When writing a technical paper some people also prefer to give the particular probability. It has already been mentioned that Yule and Kendall give the significance of χ^2 values with one degree of freedom (running 1, 1·1, 1·2, . . . up to 10) in their Table 4B. They also give t values from 0·1 to 6 (where pertinent) for up to 20 degrees of freedom in their Table 5. Normal deviates from the normal distribution can be read from Fisher and Yates' Table 1, or they can be obtained more precisely by adding 5 and using their table of probits (Table IX) which, however, gives single-sided probabilities.

One quick device for finding the individual values is to plot the figures from the appropriate table onto logarithmic probability paper. If it is found that 3 values give a straight, or practically straight, line then a particular probability value can be read off for any value that can be interpolated on this line. Alternatively, one can use formulae, and such formulae for transforming, t, F, and χ^2 values to an equivalent normal deviate are given by Wallis and Roberts in the technical notes of their Chapter 13.

SUMMARY

This chapter adds a few supplementary technical methods but, mainly, re-emphasises that the first thing to do with experimental results is to display them in a simplified form, such as frequency tables, histograms based on dot diagrams, or scatter diagrams, in order to answer the different questions that have been asked. It is the inspection of these visual aids that allows the experimenter to decide if a test of significance is required. It is indicated that the easiest test is applied first.

References

Fisher, R. A. (1938). "Statistical Methods for Research Workers." Oliver and Boyd, Edinburgh. Section 21.1, 7th Edn., page 104.

Mainland, D. (1948). "Statistical Methods in Medical Research." *Canad. J. Res.*, E 26:1-166, page 55.

—— (1952). "Elementary Medical Statistics." Saunders and Co. London, page 126.

Mosteller, F., and Tukey, J. W. (1949). "The Uses and Usefulness of Binomial Probability Paper." *J. Amer. Stat. Ass.*, **44**, 174.

Snedecor, G. W. (1957). "Statistical Methods." 5th Edn., Iowa State College, Iowa, section 9.4, page 216.

Wallis, W. A., and Roberts, H. V. (1957). "Statistics" Methuen and Co. Ltd., London. Section 19.6, example 615.

CHAPTER 14

WRITING SCIENTIFIC PAPERS

There are two classes, those who want to know, and do not care whether others think they know or not, and those who do not care very much about knowing but care very greatly about being reputed as knowing.

SAMUEL BUTLER (on scientists)

When a series of experiments is finished then, to complete the investigation, the results must be analysed and the work written up for publication in a technical journal. By this act you accept full responsibility for the work you have been paid to do, and you add the results of this work to the general pool of knowledge. It is only taking this responsibility that gives one the urge to scrutinise the results critically, for in publishing them you are stating to scientists throughout the world that by using the methods and materials you have described, they will obtain the results that you have recorded. Any failure on their part to obtain the same class of result questions your ability.

Most scientists like to publish their work in this manner but in addition to these there is, unfortunately, an increasing number who work directly for a government department or for a commercial firm. Neither of these institutions take the slightest pleasure in being criticised, nor do they always want the strength of the evidence to be known. Their scientists are therefore pressed to record their findings as reports that have a restricted circulation, here the results are bottled up and placed in the cellar instead of being added to the pool of knowledge. Thus very often in committees, government workers may refer to work they did years before which is quite unknown to the other members, and of which there is no method of judging the reliability. This is a dangerous trend for research work is becoming so expensive that perhaps, before long, only governments and rich commercial firms will be able to afford it.

Although it is important to publish one's scientific work one need not go to the other extreme and develop what is sometimes known as " literary diarrhoea." This distressing complaint is a condition in which there is an uncontrollable impulse to publish every piece of work that is carried out, whether it is good or bad, complete or incomplete. In its most acute form the same piece of work is viewed from two or three different aspects and each of these provides a paper for the journal of a different country. Infection may be due to the thought that it will bring promotion, particularly where judgement will be given by people not specialising in the same subject, or it may be due to encouragement from one's superiors where the laboratory depends on money collected from the public. Alternatively, one may become infected through ignorance of what constitutes a good paper, or from the mistaken impression that some Universities award a D.Sc. degree once published work reaches a certain weight, and are as equally impressed by trivia as they are by well planned work.

When engaged in whole-time research work one must realise that only a part of one's work will be worth publishing, and the remainder, well indexed and well annotated, must be stored as background experience that is available at any moment at which it becomes relevant.

Material for a scientific publication

Scientific progress depends on what is called " Truth by verification," a method that has stood the test of time from the days of Galileo. The hypothesis is conceived and experiments are carried out to prove or disprove it. Once the hypothesis has been proved beyond reasonable doubt the investigation is ready to be written up. Alternatively, when every reasonable effort to prove the hypothesis has failed, and the worker feels that neither he nor anybody else can succeed, then this result should be published. Braude, Kon and Porter (1953) write with feeling on this point in the following quotation: " This could be more certainly assessed if workers would publish their negative findings with the same enthusiasm as they show in publishing observations on a spectacular response." For a complete map it is as necessary to show the path that ends in a precipice as well as the one that leads to water.

It is not always possible to prove an hypothesis completely right, or completely wrong, and this has given rise to a saying that Faith is certainty without proof and Science is proof without certainty. Nevertheless the individual worker will be judged on his reliability. One thing he must avoid, and that is making a big assumption in the hypothesis, and then, after a few futile experiments, following this with a publication which ends in a statement such as " The writer has formed the opinion . . .," or " This evidence supports the author's hypothesis that . . ." These are conclusions that admit that he has failed to obtain convincing proof, and that somebody else must now do the work of proving or disproving his hypothesis.

Another type of paper is observational, in which an unusual condition, or event, is recorded as accurately as possible. Here again the account should be concise and factual, for its purpose is to add one more brick to the wall of knowledge. The temptation to load such a paper with hypotheses should be resisted, for as a factual account it will remain true for all time, but as a mixture of fact and hypothesis it will lose the author some credit each time a fresh discovery shows an hypothesis to be untrue, until finally readers may begin to doubt even the recorded facts.

Object of the Scientific Article

The primary object of the scientific article is not self-advertisement, it is to give the reader factual information that will be of help to him. By changing the word " statistician " into " research worker " this object is summed up by L. J. Holman (1938) in the following quotation: " It is not the job of the statistician to believe fervently and to persuade earnestly, but to discover, explain, and publish *facts*."

When beginning to write scientific articles one may have the impression that the reader will take a personal interest in you, and that he will be interested both in your personal difficulties and the ingenuity and perseverance that you have displayed in overcoming them, and also in the clever hypotheses you wish to bring forward to explain results that you do not undertsand. One may also have the misguided belief that he will be most interested in the data of all your negative efforts before you found the correct way of doing things, and why your clever experiment has shown that Dr. Bloggs and his co-workers are completely wrong.

Unfortunately one must acquire a more realistic view of the reader's wishes. The reader is a much more selfish person than we would like him to be. He asks for factual information, lucidly explained in simple language, and given in the form of a story. Alternatively, or in addition, he wants to be able to find those parts of the paper that interest him, without searching through the whole article.

Style

As it is the reader, rather than the writer, who is important, and as the reader likes articles that are easily understood, this is the quality that should be aimed at. Most books agree that this requires a style in which sentences are short, the words are simple, and jargon is avoided. The words most easily understood by the reader are those that he uses in conversation, and a series of unfamiliar words, each of which has to be translated inside the brain, makes reading tiring.

It is a curious fact that, although we all prefer to read articles written in simple familiar language, when we come to write, these simple words fail to appear and the brain gushes out things like "Our limited experience of this phenomenon precludes the hypothesis that this syndrome is initiated by a noxious agent." In the first draft it is advisable to push ahead, writing this stuff as it comes. With the second draft one should work for lucidity and ask oneself what is meant by such a statement. Was it meant that the condition was not due to a poison? If so this can be substituted for the pretentious original draft. Remember that, to an experienced reader, pretentiousness and self-satisfaction stick out as if they were written in headlines.

The final redrafting can be undertaken with a view to brevity, for editors will not accept articles when they see that the same information can be given in half the number of words. Very often cutting out unnecessary adjectives will both shorten the sentence and make its meaning more clear. But brevity must not be allowed at the expense of clarity.

General pattern

Every scientific article should tell a story. This seems obvious yet it is remarkable that one can occasionally read an article and end up by commenting "I wonder what he was trying to do," or something less polite. The story may be short and simple or it may be long and complicated, but in either case it should be clear to the reader what the story is, and why you moved from step to step. The data must be scrupulously true but I am not so sure that the story needs to be. Several illogical things may have been done, or things may have been done in an illogical order. Here, I think, that again you must remember that you are writing for the benefit of the reader, and, providing it makes no essential difference to the data or to the interpretation, things should be arranged so that they occur in an easily understood and logical order, and the bungled experiment, that merely showed the way to a more perfect experiment, should be omitted altogether.

One thing that must not be omitted is any result that has not conformed to the general pattern. If you have had 4 successful results and 1 unsuccessful one, then the unsuccessful one must be mentioned. It may tend to spoil the whole picture but if it is a fact it must not be omitted for it shows that there is still some factor that is not understood. You may have what you think is a very good explanation of this abnormality, if so, put this hypothesis in, but where there is no obvious explanation do not sit down and try to work out two or three feasible explanations, the reader can do that. An inconsistent result published in your article may be your salvation when you wish to refer to your work in the light of a better understanding.

One thing that must not be included is the data from a previous paper, added to boost the statistical odds, and added without making it very clear that it is work already published. Using the same evidence twice without telling anybody is so obviously wrong that it can hardly be done through ignornace.

When telling the story there is a conventional framework on which to hang your data. A very full skeleton is as follows :

Story	Heading
What it is about.	Title
Why you did it.	Introduction
How you did it.	Materials and Methods
What you did.	Plan of Experiment
What happened	Results
How this compares with other work and where it leads.	Discussion

Story	Heading
Main results in brief.	Summary
General principle acceptable from the results.	Conclusion
Who helped in one particular part of the work.	Acknowledgements
Books you consulted in building up the hypotheses and in explaining the present position.	Bibliography
Original work you referred to.	References

Articles do not require all these headings nor would all journals accept them. Often parts are combined, for example, Results and Discussion, or Summary and Conclusion, but it is as well to keep all the parts in mind when the article is being written.

In a large number of articles the foundation stone will be the basic table, and where possible some columns of this will incorporate a plan of the experiment. The object of the writer should be to include in this table as much of the results as is possible without producing a puzzle which would be more difficult to decipher than the written text.

Title

Although it forms the first part of the paper, the title should not be thought out until after the first draft is complete, for by then you should know what the article is really about. The object of the title is to describe, as concisely as possible, what the investigation was about and to display the terms under which the work will be cross-indexed in libraries. Thus the animal used might be the rabbit, the experimental interference might be the production of a stress reaction, and the measurement might be eosinophil counts of the bone-marrow. In this instance rabbit, stress reaction, and eosinophil count in bone-marrow should all appear in the title.

Introduction

The object of the introduction is to make the reader who is interested in the subject sufficiently aware of the direction of the research that is going on for him to be able to appreciate why the experiment recorded was the next logical step. Here one assumes that the reader has a knowledge of the subject at least equal to that given in current text-books, and that he can procure and read the review articles mentioned in the introduction. It is not intended that the introduction should trace the whole history of the subject.

In fiction stories the author tries to conceal the conclusion—that the half-sucked peppermint must have been left on the library table by the butler—until the last page; this is called suspense. There is no need to introduce suspense into a scientific article, and it is often of help to the reader if the introduction not only states the question that the experiment asks but also gives the answer that was obtained. Thus with a foreknowledge of the answer he can test the evidence as he comes to it.

Materials and methods

There is not enough space in journals, and the cost of printing is too high, to permit any detailed description of routine methods. All that can be recorded are facts required by the reader for judging the reliability of the experiment, and the special information that he would require if he wished to carry out a similar investigation.

If you have devised a method of cutting tissues embedded in butter by means of a chisel you may be very clever but this is of no importance to the reader unless the histological, or pathological changes you describe are different from those obtained by routine methods. Similarly if you do a routine differential count on a blood film stained with haematoxolin and eosin then this is of no interest if the reader would get the same type of result by using Leishman's stain. What he would require to know, however, would be the number of cells differentiated, so that he could estimate the error of the count. In bacteriology one may record that a plate count was carried out after incubating for 3 days at 37° C. Here it is quite unnecessary to detail the temperature because it is orthodox, but mention of the 3 days permits the reader to judge whether all the bacterial colonies would be visible by this time. He might also want to know the number of colonies on which the counts were based. Thus counts recorded as millions per ml. might, for example, be based on plates showing from 20 to 200 colonies.

Plan of experiment

The plan of a simple experiment is usually obvious from the basic table of the results. Here the general plan can be mentioned in the text and the reader can be referred to the tables of results for the details. In a complex investigation a description in the text is irksome and difficult to visualise. In these circumstances the experiments should be shown as a separate table and, where possible, the result for each part, summarised in one or two words, should be included in the final column to help the reader to get the gist of the investigation.

Results

The results should comprise the largest section of most articles and to help description they will be described under different sub-sections.

Display of Data. Whenever possible the results should be built round a basic table or a basic set of illustrations. Tables are very expensive to print and therefore the author should try to gauge, from previous copies of the journal he wishes to publish in, just how much detail is likely to be accepted.

The ideal is a basic table which records individual results for individual animals, for then the reader—if he wishes—can use the results almost as if he had done the experiment himself, and can see how they agree with his own hypotheses as well as with those of the writer. Unfortunately, where details are numerous this ideal is much too expensive and the results must be condensed. In a previous chapter the futility of recording measurements as maximum, minimum and mean was emphasised, and it was suggested that a histogram, or its tabular counterpart, the frequency table, was a much better method for, although the accuracy of the individual results was lost, the shape of the distribution was made apparent. If these results appear to be approximations of the normal curve, then recording can be shortened by describing the group as an average with the standard deviation of the distribution (not the s.d. of the mean).

Where the data to be condensed deals with a correlation, it is best shown as a scatter diagram, for here the distribution of the dots permits the reader to judge the error and to question the validity of any line, either calculated of freehand, that you have drawn to represent the theoretical relationship. The line without the dots conceals this error. It is much more important to present the visual evidence than to present a highly significant correlation coefficient without the data that permits the reader to check it visually.

Where paired measurements, say weights at different times, are few, and particularly where the times are irregularly spaced, it is quite wrong to join them up as if you knew that they could be correctly joined by straight lines, when you know no such thing. Here it is preferable to show each pair of measurements as a column rising from the base so that the distance between each column is clearly visible. This accentuates the fact that you can accept no responsibility for what might, or might not, have happened between each pair. It is a sobering thought that any two points can be joined by a straight line, and to take an over simple example, it would be quite wrong to record the weight of a newly-born calf and then its weight as a cow, and just join the two points with a straight line. Similarly the line must end at the last point measured, you must not extrapolate the line to reach some position for which you want a prediction; to do this is to change from a scientist to a soothsayer.

Beside the basic table, or figure, others can be added to present other aspects of the experiment. But there is a strict limit, varying from journal to journal, as to the amount of space that can be allowed for tables or figures. Because of this when a treatment has been given 3 or 4 trials, and the result is the same except for natural variation between animals, it is usual to save space by exhibiting one result and stating that this result is typical of the others.

Comments on the Data. The results given in the tables and figures are commented on in the text. As the paper is for the benefit of the reader these comments should help him to appreciate the story. To try to explain this let us make the strange assumption that the basic figure was the Victorian etching of a fireman rescuing a child through the window of a burning house. Comment on this picture might call attention to the strength and bravery of the fireman, comparable with the strength and destructiveness of the fire, and in contrast to the weakness and defencelessness of the small child. It would not help the reader to appreciate the picture if he was told that the fireman had 5 buttons on his tunic and that it was most interesting that 1 of them was undone, nor that accurate measurement of the window showed that it was 3 feet by 4 ft.-6 in., a most unusual size.

In a positive experiment the main results are a contrast between the experimental group and the control group, but pathological descriptions sometimes lack a direct control. Here an indirect control should at least be mentioned and a statement should be made that similar changes have not been seen in healthy animals of that age and species, with some indication of the numbers examined. In addition, if a lesion is said to be pathognomonic of a disease, a table should be given showing the number of cases of similar diseases in which the lesion was absent.

As the experiment is finished and is in the past it is spoken of in the past tense, thus one would say that " the animals *were* slaughtered." On the other hand, when referring to a table exhibited in the article it is usual to use the present tense. Thus one would say; " The results *are* shown in Table I." and go on to say that " It can be seen that the average for Group A *is* lower than that of Group B." Nevertheless, if you find a change of tense confusing, there is no reason why you should not go back to the past tense and write " The average for Group A *was* lower . . ." for that is equally true, and many writers prefer to do this.

When describing a figure or illustration the position is the same, so that histological work accompanied by a photomicrograph calls for the present tense. At the time of writing there is also an affectation by which all histological lesions are described in the present tense even where there is no photograph. Thus one may be told that an animal *was* slaughtered, that its liver *showed* pale patches, and that fibroblasts *are* proliferating rapidly in these spots. This in an animal that may have been dead some months! In such circumstances readers who are accustomed to use their own judgement will probably do so, but those who feel that using the past tense may imperil their reputation as pathologists should restrict themselves to using the present tense until this fashion passes.

Extra Side Headings. Also included in the results should be the things that went wrong; for the reader's sake these should be given under sectional side headings so that they can be found easily and do not muddle his understanding of the main line of the investigation. Such a side heading might cover one trial spoilt by some intercurrent infection, providing that the trial does provide information of interest to the reader. Similarly, some mutation in the organism may occur and the characteristics of this new strain might be thought worth recording. In the same way results occurring through what is almost certainly a mistake can be dealt with under a side heading. For example, a mistakenly small or large test dose, or a filtered culture that still contained bacteria; if left incorporated with the successful trials these accidents would necessitate the frequent use of statements such as "with the exception of experiment 4 in which we believe that the filtered material was contaminated" which is apt to become annoying to the writer and boring to the reader.

Discussion

This is the most enjoyable part of the paper to write. It is here that you extend your hypothesis and show its wider possibilities. Here you can explain why other people's efforts failed where yours succeeded and how, by accepting your methods, the country can save at least £5,500,000 per year. And this is where you show the reader how clearly you can think. The only disadvantage is that when you hand the draft to a senior worker for criticism it all gets crossed out.

To save yourself this disappointment the discussion must be kept to essentials. If you need to predict how much money will be saved by stopping animals dying from one disease, bear in mind that unless they are slaughtered by the butcher, they will all die of some other disease. Giving an estimate gives the reader an opportunity of judging your reliability, so be cautious.

Be chary of proclaiming other workers as wrong, instead seek for the variation in circumstances that has caused the difference in results. It may be your material that is deficient in some active principle, or your organism that has mutated. The fact that you sincerely believe that the other workers are mentally deficient must be concealed absolutely, otherwise it may be you who suffers the subsequent embarrassment, for they may turn out to be right in this particular instance.

Avoid arm-chair logic; this holds a high place in religion but a low one in science. The conjurer is the most logical man alive. He proves that the hat is empty, he shows you both sides of the cloth and then, after touching the covered hat with his magic wand, produces a live rabbit which it is impossible to doubt. If you believe in arm-chair logic you will be convinced that he really possesses a magic wand, but if you are a realist you may feel that half-an-hour testing the apparatus will help you much more than hours of arm-chair thinking. Remember that a chain of reasoning is quite useless unless you are sure that each of the facts is correct. Do not spoil a reliable paper by speculating on what, at that moment, are merely believed to be the facts. Finally do not burden the discussion with a list of all the alternative hypotheses that come to mind, in the hope that when one of them is found to be correct you can say that you thought of it first.

Summary

This is an important part of the paper because it is often the only part that the reader consults. Because of this some journals are now putting the summary at the beginning of the article. Some writers appear to resent the fact that the reader might try to avoid reading the entire article and make the summary as uninformative as possible. Thus, if describing the preparation of marmalade they would write: "This article describes the preparation of marmalade. It details the types of fruit required and the method by which it is prepared." This is equivalent to saying "I've had a lot of work writing this article, if you want any real information out of it you can jolly well read it!" If we accept that the reader is the important person, then we try to give a more specific summary and might begin: "Marmalade was prepared by slicing 2 lemons and 12 oranges. After pips had been removed . . . ."

Conclusion

In many journals the conclusion is often given as the last line of the summary; others permit a separate conclusion. The conclusion is a statement such as : "The injection of tablets of thyroxin increased the wool yield of sheep." If you were certain of this correlation you could go farther and predict for the future by putting it in the present tense, thus: "The injection of thyroxin *increases* the wool yield in sheep." On the other hand if you wish to qualify the statement because you believe that the increase only occurs when the sheep are on a poor diet, you can say: "Under the conditions of the experiment the injection . . . etc."

Acknowledgements

Acknowledgements tend to be emotional, thus many junior workers wish to write an acknowledgement beginning: "This work was done and this paper is published despite every discouragement from . . ." There are no rules for acknowledgements and their contents change with time and vary with the laboratory. As a guide it can be said that they should mention anyone who has given you help or material without being paid for it. On the other hand it is courteous to show an acknowledgement to the people concerned before publication, for some object to fulsome acknowledgement and others object to being thanked for advice in a paper that describes what they believe to be badly planned experimental work.

References

References should be put in according to the method asked for by the journal to which you are sending the contribution. The important point is to check them very carefully before publication, for it is hard to convince a reader that you are an accurate and reliable worker if he has the misfortune to find that the first reference he selects is not the correct guide to the article he wishes to consult.

Criticism

Having completed the article it should be passed round for criticism. Nobody likes having his work pulled to pieces but it is far better to have it done by a few fellow workers before the paper is published than by the majority of readers after publication, when it is too late to alter it. Nor need one worry too much about the quality of the scrutineers for it is surprising how people who have never carried out a well-planned sensible experiment in their lives can give a useful and objective criticism when reading someone else's work. Thus it is very unusual for a paper to be passed round 3 or 4 people without the author receiving suggestions that would improve the clarity or the presentation.

Further reading

Nuttall wrote a small book on the preparation of scientific papers and although I think it is out of print it will be found in most scientific libraries. More recently Jolly (1958) has written 2 or 3 pages, in lighter vein, on the subject. To help in the use of simple English there are 2 small booklets entitled *Plain Words* and *The A.B.C. of Plain Words*, by Sir Ernest Gowers. For those readers who wish for detailed instruction in writing scientific books or papers there are two American books that can be recommended. These are by Hewitt (1957) and by Emberger and Hall (1955).

Summary

From the editor's point of view, scientific articles are not published just to advertise the writer, nor to enable him to get rid of the data he has acquired during the year. They are intended for the interest and education of the reader, and should be written from that point of view.

The article should tell a simple story, such as, the facts leading to an hypothesis, how this hypothesis was tested, and how this test proved, or disproved, the theory. The value of the article is seldom increased by the inclusion of several additional hypotheses.

References

Braude, R., Kon, S. K., and Porter, J. W. G. (1953). "Antibiotics in Nutrition." *Nutr. Abstr. Rev.*, **23**, 473.

Emberger, M. R., and Hall, M. R. (1955). "Scientific Writing." Harcourt, Bruce and Co., N.Y.

Gowers, E. (1954). "The Complete Plain Words." Penguin Books, Ltd., Middlesex.

Hewitt, R. M. (1957). "The Physician-Writer's Book." W. B. Saunders.

Holman, L. J. (1938). "Simplified Statistics." Pitman and Sons, London page 131.

Jolly, D. W. (1958). "This Writing Business." *Vet. Rec.* **70**, 983

APPENDIX

Table III (Mainland, Herrera and Sutcliffe)

Probabilities for fourfold contingency tables.

N = total in each sample. Probabilities from one tail are in parentheses.

N	No. of A's in Sample (1)/No. of A's in Sample (2).					
10	0/10(·0000)	0/9(·0001)	0/8(·0004)	0/7(·0015)	0/6(·0054)	0/5(·0163)
	0/4 (·0433)	0/3(·1053)	0/2(·2369)	0/1(·5000)	1/9(·0005)	1/8(·0027)
	1/7 (·0099)	1/6(·0286)	1/5(·0704)	1/4(·1517)	1/3(·2910)	1/2(·5000)
	2/8 (·0115)	2/7(·0349)	2/6(·0849)	2/5(·1749)	2/4(·3143)	2/3(·5000)
	3/7 (·0894)	3/6(·1849)	3/5(·3250)	3/4(·5000)	4/6(·3281)	4/5(·5000)

Abridged Table IV (Fisher and Yates)

Distribution of χ^2

n	·80	·70	·50	·30	·20	·10	·05	·02	·01	·001
1	·064	·148	·455	1·07	1·64	2·71	3·84	5·41	6·64	10·8
2	·446	·713	1·39	2·41	3·22	4·61	5·99	7·82	9·21	13·8
3	1·00	1·42	2·37	3·66	4·64	6·25	7·82	9·84	11·3	16·3
4	1·65	2·20	3·36	4·88	5·99	7·78	9·49	11·7	13·3	18·5
5	2·34	3·00	4·35	6·06	7·29	9·24	11·1	13·4	15·1	20·5
6	3·07	3·83	5·35	7·23	8·56	10·6	12·6	15·0	16·8	22·5
7	3·82	4·67	6·35	8·38	9·80	12·0	14·1	16·6	18·5	24·3
8	4·59	5·53	7·34	9·52	11·0	13·4	15·5	18·2	20·1	26·1
9	5·38	6·39	8·34	10·7	12·2	14·7	16·9	19·7	21·7	27·9
10	6·18	7·27	9·34	11·8	13·4	16·0	18·3	21·2	23·2	29·6
11	6·99	8·15	10·3	12·9	14·6	17·3	19·7	22·6	24·7	31·3
12	7·81	9·03	11·3	14·0	15·8	18·5	21·0	24·0	26·2	32·9
13	8·63	9·93	12·3	15·1	17·0	19·8	22·4	25·5	27·7	34·5
14	9·47	10·8	13·3	16·2	18·2	21·1	23·7	26·9	29·1	36·1
15	10·3	11·7	14·3	17·3	19·3	22·3	25·0	28·3	30·6	37·7
16	11·2	12·6	15·3	18·4	20·5	23·5	26·3	29·6	32·0	39·3
20	14·6	16·3	19·3	22·8	25·0	28·4	31·4	35·0	37·6	45·3
25	18·9	20·9	24·3	28·2	30·7	34·4	37·7	41·6	44·3	52·6
30	23·4	25·5	29·3	33·5	36·2	40·3	43·8	48·0	50·9	59·7

Abridged Table III (Fisher and Yates)

Distribution of t

n	·7	·6	·5	·4	·3	·2	·1	·05	·02	·01	·001
1	·51	·73	1·00	1·38	1·96	3·08	6·31	12·7	31·8	63·7	637
2	·44	·62	·82	1·06	1·39	1·89	2·92	4·30	6·96	9·92	31·6
3	·42	·58	·76	·98	1·25	1·64	2·35	3·18	4·54	5·84	12·9
4	·41	·57	·74	·94	1·19	1·53	2·13	2·78	3·75	4·60	8·61
5	·41	·56	·73	·92	1·16	1·48	2·02	2·57	3·36	4·03	6·86
6	·40	·55	·72	·91	1·13	1·44	1·94	2·45	3·14	3·71	5·96
7	·40	·55	·71	·90	1·12	1·42	1·90	2·36	3·00	3·50	5·40
8	·40	·55	·71	·89	1·11	1·40	1·86	2·31	2·90	3·35	5·04
9	·40	·54	·70	·88	1·10	1·38	1·83	2·26	2·82	3·25	4·78
10	·40	·54	·70	·88	1·09	1·37	1·81	2·23	2·76	3·17	4·59
11	·40	·54	·70	·88	1·09	1·36	1·80	2·20	2·72	3·11	4·44
13	·39	·54	·69	·87	1·08	1·35	1·77	2·16	2·65	3·01	4·22
15	·39	·54	·69	·87	1·07	1·34	1·75	2·13	2·60	2·95	4·07
20	·39	·53	·69	·86	1·06	1·32	1·72	2·09	2·53	2·84	3·85
25	·39	·53	·68	·86	1·06	1·32	1·71	2·06	2·48	2·79	3·72
30	·39	·53	·68	·85	1·06	1·31	1·70	2·04	2·46	2·75	3·65
60	·39	·53	·68	·85	1·05	1·30	1·67	2·00	2·39	2·66	3·46

Abridged Table V (Fisher and Yates)

Variance Ratio. Significant where p = ·05.

Larger variance

n^2 / n^1	1	2	3	4	6	8	12
1	161	199	215	225	234	239	244
2	18·5	19·0	19·2	19·3	19·3	19·4	19·4
3	10·1	9·6	9·3	9·1	8·9	8·8	8·7
4	7·7	6·9	6·6	6·4	6·2	6·0	5·9
5	6·6	5·8	5·4	5·2	4·9	4·8	4·7
6	6·0	5·1	4·8	4·5	4·3	4·2	4·0
7	5·6	4·7	4·3	4·1	3·9	3·7	3·6
8	5·3	4·5	4·1	3·8	3·6	3·4	3·3
9	5·1	4·3	3·9	3·6	3·4	3·2	3·1
10	4·9	4·1	3·7	3·5	3·2	3·1	2·9
12	4·7	3·9	3·5	3·3	3·0	2·8	2·7
14	4·6	3·7	3·3	3·1	2·8	2·7	2·5
16	4·5	3·6	3·2	3·0	2·7	2·6	2·4
18	4·4	3·6	3·2	2·9	2·7	2·5	2·3
20	4·4	3·5	3·1	2·9	2·6	2·4	2·3
25	4·2	3·4	3·0	2·8	2·5	2·3	2·2
30	4·2	3·3	2·9	2·7	2·4	2·3	2·1
40	4·1	3·2	2·8	2·6	2·3	2·2	2·0

ABRIDGED TABLE OF WHITE'S RANKING TEST

n_2 ↓ / n_1 →	2	3	4	5	6	7	8	9	10	11	12	13	14	15
$p = 0{\cdot}05$														
4			10											
5		6	11	17										
6		7	12	18	26									
7		7	13	20	27	36								
8	3	8	14	21	29	38	49							
9	3	8	15	22	31	40	51	63						
10	3	9	15	23	32	42	53	65	78					
11	4	9	16	24	34	44	55	68	81	96				
12	4	10	17	26	35	46	58	71	85	99	115			
13	4	10	18	27	37	48	60	73	88	103	119	137		
14	4	11	19	28	38	50	63	76	91	106	123	141	160	
15	4	11	20	29	40	52	65	79	94	110	127	145	164	185
16	4	12	21	31	42	54	67	82	97	114	131	150	169	
17	5	12	21	32	43	56	70	84	100	117	135	154		
18	5	13	22	33	45	58	72	87	103	121	139			
19	5	13	23	34	46	60	74	90	107	124				
20	5	14	24	35	48	62	77	93	110					
$p = 0{\cdot}01$														
5				15										
6			10	16	23									
7			10	17	24	32								
8			11	17	25	34	43							
9		6	11	18	26	35	45	56						
10		6	12	19	27	37	47	58	71					
11		6	12	20	28	38	49	61	74	87				
12		7	13	21	30	40	51	63	76	90	106			
13		7	14	22	31	41	53	65	79	93	109	125		
14		7	14	22	32	43	54	67	81	96	112	129	147	
15		8	15	23	33	44	56	70	84	99	115	133	151	171
16		8	15	24	34	46	58	72	86	102	119	137	155	
17		8	16	25	36	47	60	74	89	105	122	140		
18		8	16	26	37	49	62	76	92	108	125			
19	3	9	17	27	38	50	64	78	94	111				
20	3	9	18	28	39	52	66	81	97					
$p = 0{\cdot}001$														
7						28								
8					21	29	38							
9				15	22	30	40	50						
10				15	23	31	41	52	63					
11				16	23	32	42	53	65	78				
12				16	24	33	43	55	67	81	95			
13			10	17	25	34	45	56	69	83	98	114		
14			10	17	26	35	46	58	71	85	100	116	134	
15			10	18	26	36	47	60	73	87	103	119	137	156
16			11	18	27	37	49	61	75	90	105	122	140	
17			11	19	28	38	50	63	77	92	108	125		
18			11	19	29	39	51	65	79	94	111			
19			12	20	29	41	53	66	81	97				
20			12	20	30	42	54	68	83					
21		6	12	21	31	43	56	70						

Dr Colin White (1952). *Biometrics*, **8**, 33.
Reproduced by the kind permission of the author and editors.

SELECTED BIBLIOGRAPHY

MATHEMATICS

COMRIE, L. J. *Barlow's Tables.* E. and F. N. Spon, Ltd., London. Includes squares, square-roots, reciprocals, binomial coefficients, and Bessel's method of interpolation.

FELDMAN, W. M. *Biomathematics.* Chas. Griffin and Co. Ltd., London. A useful book on mathematics, now out of print but in most scientific libraries. A new edition by C. A. B. Smith makes less allowance for non-mathematicians.

SAWYER, W. W. *Mathematician's Delight.* Penguin Books, Ltd., Middlesex. A delightful book on how to study mathematics.

WESLEY, R. *Mathematics for All.* Odham's Press, London. Revises simple mathematics up to elementary calculus in 132 pages, and teaches its application to domestic and scientific subjects, including biology, games of chance and statistics.

INTRODUCTION TO STATISTICS

HILL, A. B. "Principles of Medical Statistics." *Lancet,* London. Taught from the author's concept of statistics as "arithmetic guided by logic." Emphasis on avoiding fallacies.

LEVY, H., and PREIDEL, E. E. *Elementary Statistics.* Thos. Nelson and Sons, Ltd., London. A short, introductory course to the mathematical basis of statistical tests by expert teachers. Worth consulting from time to time.

MAINLAND, D. *Treatment of Clinical and Laboratory Data.* Oliver and Boyd Ltd., Edinburgh and London. A very simple introduction to applied statistics for the non-mathematician. Covers most techniques required by the junior worker.

APPLIED STATISTICS

FISHER, R. A., and YATES, F. *Statistical Tables for Biological, Agricultural and Medical Research.* Oliver and Boyd Ltd., Edinburgh and London. An almost essential book for the library.

MAINLAND, D., HERRERA, L., and SUTCLIFFE, MARION I. *Tables for Use with Binomial Samples.* Dept. Med. Stat., New York Univ. Col. of Med., 550 First Av., N.Y. 16. Tables of significance based on the exact test, the χ^2 test, and binomial confidence levels, by which results can be tested in a few seconds. (Order direct.)

MATHER, K. *Statistical Analysis in Biology.* Methuen and Co. Ltd., London. Although a small book it contains well selected examples which cover most experiments and, if later you become interested in the mathematical side, the demonstrations are there.

MORONEY, M. J. *Facts from Figures.* Penguin Books, Ltd., Middlesex. This useful and amusing book includes a very clear exposition of the analysis of variance. It is worth having in addition to any other books.

SNEDECOR, G. W. *Statistical Methods Applied to Experiments in Agriculture and Biology.* Iowa State Col. Press, Iowa. This popular book covers the routine requirements of biologists and is widely used.

SPECIALISED STATISTICS

FINNEY, D. J. *Probit Analysis.* Cambridge Univ. Press, Cambridge. Except for the introductory chapters, this is for specialists only.

YATES, F. *The Design and Analysis of Factorial Experiments.* Imperial Bureau of Soil Science, Harpenden, U.K. An inexpensive booklet dealing with different factorial designs, including split-plot experiments.

YATES, F. *Sampling Methods for Censuses and Surveys.* Chas. Griffin and Co. Ltd., London. Gives simple description for non-specialists on how to carry out surveys, followed by technical descriptions of the different designs and their analysis.

YULE, G. U., and KENDALL, M. G. *An Introduction to the Theory of Statistics.* Chas. Griffin and Co. Ltd., London. A popular text-book for those interested in the theoretical basis of statistics.

PHILOSOPHY OF RESEARCH

ARBER, AGNES. *The Mind and the Eye.* Cambridge Univ. Press, Cambridge. The logic and philosophy of biological research method.

BEVERIDGE, W. I. B. *The Art of Scientific Investigation.* W. Heineman Ltd., London. A stimulating book illustrated by many examples from the work of famous men.

WRITING SCIENTIFIC PAPERS

EMBERGER, M. R., and HALL, M. R. *Scientific Writing.* Harcourt, Brace and Co., New York. A text-book on the routine of preparing and writing of scientific books and articles.

NUTTALL, G. H. F. *Notes on the Preparation of Papers for Publication.* Cambridge Univ. Press, Cambridge. Concise information and advice on writing scientific articles. Out of print but in most scientific libraries.

STANILAND, L. N. *The Principles of Line Illustration.* Burke, London. Techniques of lettering and producing diagrams and illustrations for scientific publications.

INDEX